站在巨人的肩上
Standing on Shoulders of Giants

TURING
图灵教育

iTuring.cn

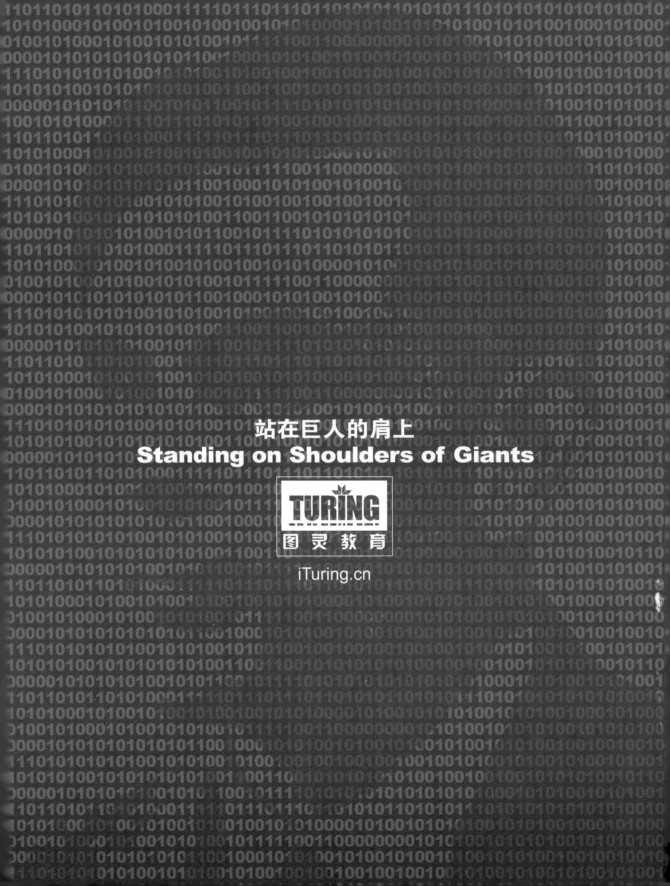

图灵原创

王利华 魏晓军 冯诚祺◎编著

React Native
入门与实战

人民邮电出版社

北京

图书在版编目（CIP）数据

　　React Native入门与实战 / 王利华，魏晓军，冯诚祺编著. -- 北京：人民邮电出版社，2016.1（2016.4重印）
　　（图灵原创）
　　ISBN 978-7-115-41191-4

　　Ⅰ．①R… Ⅱ．①王… ②魏… ③冯… Ⅲ．①移动终端－应用程序－程序设计 Ⅳ．①TN929.53

　　中国版本图书馆CIP数据核字(2015)第290196号

内 容 提 要

　　本书共4部分，首先简要介绍了开发相关的基础知识，然后介绍了React Native 的API、组件以及Native 扩展和组件的封装，接着介绍了App 的动态更新和上架过程，最后通过3 个案例介绍了如何使用React Native 开发原生App。

　　本书适合想使用React Native 开发原生应用的人阅读。

◆ 编　　著　王利华　魏晓军　冯诚祺
　　责任编辑　王军花
　　责任印制　杨林杰

◆ 人民邮电出版社出版发行　北京市丰台区成寿寺路11号
　　邮编　100164　电子邮件　315@ptpress.com.cn
　　网址　http://www.ptpress.com.cn
　　三河市海波印务有限公司印刷

◆ 开本：800×1000　1/16
　　印张：24.25
　　字数：454千字　　　　　　　2016年 1 月第 1 版
　　印数：7 001 – 10 000册　　　2016年 4 月河北第 3 次印刷

定价：79.00元
读者服务热线：(010)51095186转600　印装质量热线：(010)81055316
反盗版热线：(010)81055315
广告经营许可证：京东工商广字第 8052 号

序 一

When I heard React Native, It was 2014-8 from a Facebook friend who was in the team. They were creating this new technology for Facebook App (internal use). He said Facebook plans to open source this technology to App developers. In 2015-3, React Native was released to the public.

Although the technology have limitations across iOS and Android, its design approach was revolutionary. It proves to work close to native App experience (performance) as well as to leverage web developers skills. React Native has already impact on App developers now. It becomes a top choice for developing new App.

What Lihua、Xiaojun and Chengqi surprised me is they were writing this book together in last 6 months. While they are trying out React Native, they love it. They believe that this technology is useful for App development. By creating the book in Chinese, they take React Native to next level of the developer reach and make the wide impact on Chinese App Community.

Their spirit of spreading the new technology and their endeavor of creating the book together as team work should highly praised. They Not only master this technology for use by themselves, but also show it to Chinese developers. The mobile internet is pushing forward , with many developers in this spirit and effort!

2014年8月，我从一个在Facebook工作的朋友那里听说了React Native，他们正在使用这种新技术开发内部App。他表示，Facebook计划将该技术开源。2015年3月，React Native公开发布了。

尽管这项技术在兼容iOS和Android方面仍有一定的局限性，但它的设计理念别出心裁。在实现媲美原生 App 的用户体验（性能）的同时，React Native允许Web开发者更多地基于现有经验开发App。现在，它已经影响了很多App开发者，在开发下一个App时，React Native将成为开发人员的首选技术。

在过去的6个月中，利华、晓军和诚祺合力完成了本书的创作，这让我感到惊喜。他们在不断尝试 React Native，他们热爱这项技术。他们坚信，这门技术对于 App 开发者而言十分有帮助。《React Native 入门与实战》一书的出版，将使更多开发人员接触到这项技术，并对国内 App 开发者社区产生重要影响。

在创作本书的过程中，他们的分享精神和团队协作都值得表扬。他们不仅自己掌握了这门技术，同时也将其推介给了所有国内的开发人员。移动互联网的发展，无疑得益于众多像他们这样的开发人员的努力贡献和分享精神。

2015-11-20

Eric Ye，携程旅行网CTO

序　二

随着苹果公司推出iPhone和App Store，移动开始持续升温，各种创新不断，风头逐渐盖过Web，吸引了大量开发人员进入移动领域。这无疑使得苹果公司成为最大的赢家，操控着很多公司的命运。

Facebook在Web技术上非常成功，深知移动未来的重要性，但又不想受制于苹果公司。于是投入大量的人力和物力，在移动HTML5上攻坚克难，虽取得了不少进展，但始终不如意。2012年9月，Facebook表示："Betting on HTML5 was a mistake."全力转型Native App开发。令人没有想到的是，两年之后Facebook居然推出了React Native for iOS技术，让人眼前一亮，兴奋不已。

大家都知道，Native App因其性能优越和功能强大而笑傲江湖，但终究逃不出Apple的掌心，多版本维护非常痛苦。HTML5虽然有Web的优势，但因WebView在移动设备的性能和电力等因素的制约，性能总被诟病，难成大器。而Hybrid App集Native App和Web优点于一体，还可以相互补短，似乎应该成为大家的选择。然而它在成熟度、标准化等方面的顾虑，也会是一个不小的问题。React Native技术的诞生则普遍被大家接受，各大公司纷纷介入，给人很大的想象空间。它的底层引擎是JavaScript Core，调用的是原生组件而非HTML5组件（HTML+CSS+JavaScript构建的组件）。运行时，可以做到与Native App媲美的体验，同时因为JavaScript代码可以使用后端强大的Web方式管理，既可以做到高效开发，也可以做到快速部署和问题热修复。React Native App运行在客户的手机上，而控制端可以在后端，可以充分发挥Web的能力，就像一个牵线木偶，任凭你表演。

该书不仅对React Native基础知识讲解得很透彻，同时还辅助了不少案例，比如第9章的LBS应用开发和第10章的OpenAPI应用开发。相信阅读本书后，你一定会有所收获。

雄关漫道真如铁，而今迈步从头越。很高兴看到携程框架团队同学在前端框架和React Native方面所做出的努力，也希望读者进入React Native领域后能够不断丰富社区，促进React Native技术的蓬勃发展。

吴其敏，携程旅行网框架研发部负责人，高级研发总监

序 三

最近三四年间，国内外的前端与全栈开发者社区都在坚持不懈地追寻使用JavaScript与HTML、CSS技术体系开发App内场景的核心工程技术。这种技术，在国内很多公司与团队中，被通称为H5。这种工程类的尝试最早出现在新闻资讯页等强排版、弱交互的产品场景中，因其灵活的布局能力和免发版的敏捷迭代潜力而大受欢迎。而后在涌现出大批第三方应用市场的浪潮中，也成为了应用市场展示App应用详情的技术标配。在此过程中，PhoneGap（后来的Apache Cordova）等组件的出现满足了JavaScript与Android/iOS程序之间的通信需求，及时补全了这种工程方案在系统能力上的短板。大家在更大范围地推进这种方案的过程中，却遇到了一个致命的问题，那就是这种技术在处理无限滑动列表（如微博的信息流）时，受WebView的影响，表现出了极差的点击响应、内存性能和兼容性。社区中有不少有识之士提出了模板配置化+原生渲染，或引入多个WebView运行SPA以缓解内存问题等行之有效的方案，但均因一定程度上牺牲了灵活、敏捷的方案优势，而无法获得广泛采纳。这样的缺陷直接制约了JavaScript in App工程方案的再扩大。在百度供职时，我曾负责主持Clouda云端一体框架的研发工作，可说是在社区一线全程参与了这个演进过程。

React Native的出现，彻底、整洁且智慧地解决了这个痛点，且通过WebKit的引入完整保留了JavaScript语言的完整逻辑性，通过原生的渲染保持了"不沾手"的顺滑体验和出色的内存管理，没有妥协地实现了大家需要的JavaScript in App的工程体系能力。

本书的第一作者利华曾与我在阿里高德共事，当时就表现出了对社区技术发展的很强敏感性。本书的出版，距React Native宣布支持Android平台仅月余，可以说是React Native实战的第一手中文资料。

我推荐大家阅读这本书。在掩卷之后，你当能执React Native于实际工程项目中快速作战，甚至实践于"云控App"的高精尖领域。届时，我也希望读者你进一步思考与了解React Native的设计与结构，在社区中贡献一份属于你的力量，为这个领域中更多的同学引路。

童遥，阿里高德开放平台总经理

序　四

从React Native诞生那一天起，对于作为后端出身但一直对前端念念不忘的我来说，就像发现了一个埋藏很深亟待挖掘的金矿一样兴奋。

可以说，从16年前第一次接触DHTML后，只有NodeJS的出现，才能第二次让我产生对前端技术的兴奋感。而这一次，React Native的横空出世，再次触及了我快要冷淡下去的前端兴奋感。

本书作者魏晓军是我在携程的同事，也是当时及现在携程技术团队中的顶尖前端高手。晓军所领导的携程H5/Hybrid框架团队，是整个携程移动优先战略得以顺利推进的核心技术动力。

感谢晓军和他的团队为前端技术社区及广大前端技术爱好者推出这本期待已久的作品，而我，作为一名仍对前端技术念念不忘的 IT 从业人员，也会在该书上市后第一时间抢购回家，饱饱眼福。

张雪峰，饿了么CTO，于饿了么总部

致 谢

因为我们团队比较早关注了React Native，同时也做了些分享，因此有幸结识了图灵公司的王军花老师。2015年6月24日，收到出版社王老师的邮件，希望我们能写一本关于React Native的图书，但是因为React Native才刚刚发布不久，所以在思考了一段时间后，才下定决心写作。

新事物的发展总是曲折的。React Native就像一个新生儿一样接受着质疑和考究。我们期待它更加完善、更加稳定。在本书的写作过程中，我们一直关注着React Native的更新和发展，本书的内容也随着React Native的发展而修改了多次。

对于工程师而言，写作是一件比较痛苦的事，我们更加喜欢使用代码来表达我们的想法。很多时候，一句话需要推敲好多次才能下笔。其中，初稿有很多不完善的地方，都是王老师日夜兼程地提出建议和修改，这里要对王老师表示深深的感谢！写作本书，对我们而言，不仅加深了对React Native的理解，同时也提升了个人的写作水平，这要感谢人民邮电出版社图灵公司给予的机会。

本书得以完成要感谢很多人，感谢储诚栋等领导对工作的支持和对本书写作的指导，感谢Eric Ye、吴其敏、童遥、张雪峰为本书慷慨作序，感谢储诚栋、陈浩然、勾股、鬼道为本书写推荐语，感谢潘菲菲、卢玮、蒋竟等同事的帮助。

人生难得3万日，青春区区5千天。希望本书能成为React Native开发者的小助手，也希望我们和学习React Native的同学一起努力，为React Native的蓬勃发展贡献自己的力量。

王利华　魏晓军　冯诚祺
2015年11月于上海

前　　言

React Native开启了开发原生App的新方式，不仅提高了开发效率，同时提高了App的用户体验。相比Web App而言，React Native可以使用原生的组件和API，这样就可以释放Native的能力和体验；相比Native开发而言，前端开发者可以使用JavaScript开发原生应用，这样开发效率将会得到很大的提高。

本书目的

目前，国内针对React Native讲解的图书和资料都很少，阅读本书可以帮助你更好地开发React Native应用。或许你已经了解React Native的基本内容，或许你已经开始了React Native的开发之旅，无论如何，本书都希望可以带领大家拥抱React Native，使用React Native。当然，本书也希望弥补中文资料在这方面的欠缺。

内容和组织结构

本书的内容是我们在实践过程中总结得到的，一共分为4部分。

第一部分为基础语法篇，共两章内容，主要介绍了React Native的开发基础知识。

第1章介绍了React Native的环境搭建、React与React Native之间的关系，以及如何学习React Native。

第2章主要介绍了React Native的开发基础知识，包含flexbox布局、JSX语法，并且详细介绍了React Native创建项目的过程。

第二部分为API和组件篇，共4章内容，主要介绍了React Native的API、组件以及Native扩展和组件的封装。

第3章介绍了React Native常用组件，包含View组件、Text组件、NavigatorIOS组件、TextInput组件、Touchable类组件（TouchableHighlight、TouchableOpacity和TouchableWithoutFeedback）、Image组件、TabBarIOS组件和WebView组件。

第4章介绍了React Native常用的API，包含AppRegistry、AsyncStorage、AlertIOS、ActionSheetIOS、PixelRatio、AppStateIOS、StatusBarIOS、NetInfo、CameraRoll、VibrationIOS、Geolocation、网

络数据请求的应用以及定时器和动画。

第5章介绍了React Native的实现原理以及在原生组件和API上的扩展，并且以一个"图表"组件为案例进行实战讲解。

第6章介绍了使用JavaScript封装React Native组件，主要实现了二级菜单组件、日历组件以及初步介绍了开源组件的用法。

第三部分为App更新和上架篇，共一章内容。

第7章介绍了App的动态更新和上架过程。

第四部分为实战篇，共3章内容，介绍了如何使用React Native开发原生App。

第8章介绍了使用React Native和Node.js开发企业内部通讯录应用——"百灵鸟"App。

第9章介绍了使用React Native Geolocation API和高德地图API开发LBS应用——"附近"App。

第10章介绍了使用豆瓣开放API开发一款搜索App，主要包含图书、电影和音乐搜索。

本书特色介绍

本书的特色主要在于理论结合实战，读者不仅可以了解React Native的API和组件，同时可以通过案例和实战深入学习。

源代码

本书包含的代码及其案例可以到https://github.com/vczero/React-Native-Code下载或者到图灵社区本书主页免费注册下载。本书创作时间较短，难免会有疏漏，恳请各位读者斧正。

王利华

2015年11月于上海凌空SOHO

目 录

第一部分　基础语法篇

第1章　React Native 简介 ·················· 2
1.1　环境搭建 ·· 2
　　1.1.1　安装 Node.js ······························ 2
　　1.1.2　安装 React Native ······················ 6
　　1.1.3　使用 NVM 管理 Node.js 版本 ···· 7
　　1.1.4　创建项目 ·································· 8
1.2　从 React 到 React Native ····················· 9
　　1.2.1　React 简介 ································ 9
　　1.2.2　React Native 简介 ··················· 13
1.3　为什么要使用 React Native ··············· 17
1.4　如何学习 React Native ······················ 17
1.5　说明 ··· 18

第2章　React Native 开发基础 ········· 19
2.1　flexbox 布局 ····································· 19
　　2.1.1　介绍 ··· 19
　　2.1.2　布局模型 ·································· 20
　　2.1.3　伸缩容器属性 ·························· 20
　　2.1.4　伸缩项目属性 ·························· 37
　　2.1.5　在 React Native 中使用
　　　　　flexbox ···································· 46
　　2.1.6　实例 ··· 47
2.2　React 中的 JSX ·································· 53
　　2.2.1　JSX 入门 ·································· 53
　　2.2.2　JSX 实战之 ReactJS ·················· 59
　　2.2.3　JSX 实战之 React Native ·········· 77

2.3　React Native 开发向导 ······················ 80
　　2.3.1　配置文件 ·································· 80
　　2.3.2　运行 ··· 83
　　2.3.3　调试 ··· 85
　　2.3.4　内部发布 ·································· 91
2.4　参考资料 ·· 91

第二部分　API 和组件篇

第3章　常用组件及其实践 ················· 94
3.1　View 组件 ··· 94
　　3.1.1　View 介绍 ································ 94
　　3.1.2　案例：九宫格实现 ··················· 94
3.2　Text 组件 ·· 102
　　3.2.1　Text 组件介绍 ························ 102
　　3.2.2　案例：网易新闻列表展示 ······ 102
3.3　NavigatorIOS 组件 ·························· 109
　　3.3.1　NavigatorIOS 组件介绍 ·········· 109
　　3.3.2　案例：列表页跳转详情页 ······ 110
3.4　TextInput 组件 ································ 113
　　3.4.1　TextInput 组件介绍 ················ 113
　　3.4.2　案例：搜索自动提示 ············· 114
3.5　Touchable 类组件 ··························· 121
　　3.5.1　TouchableHighlight 组件 ········ 121
　　3.5.2　TouchableOpacity 组件 ·········· 123
　　3.5.3　TouchableWithoutFeedback
　　　　　组件 ·· 123
3.6　Image 组件 ····································· 124

目 录

- 3.6.1 Image 组件介绍 124
- 3.6.2 加载网络图片 124
- 3.6.3 加载本地图片 127
- 3.7 TabBarIOS 组件 128
  - 3.7.1 TabBarIOS 组件介绍 128
  - 3.7.2 案例：类 QQ Tab 切换 129
- 3.8 WebView 组件 133
  - 3.8.1 WebView 组件介绍 133
  - 3.8.2 案例：使用 WebView 组件加载微博页面 134
  - 3.8.3 案例：新浪微博 OAuth 认证 136

第 4 章 常用 API 及其实践 138

- 4.1 AppRegistry 138
  - 4.1.1 AppRegistry 介绍 138
  - 4.1.2 AppRegistry 示例 138
- 4.2 AsyncStorage 140
  - 4.2.1 AsyncStorage 介绍 140
  - 4.2.2 案例：购物车 140
- 4.3 AlertIOS 149
  - 4.3.1 AlertIOS 149
  - 4.3.2 AlertIOS 组件的应用 150
- 4.4 ActionSheetIOS 152
  - 4.4.1 ActionSheetIOS 介绍 152
  - 4.4.2 ActionSheetIOS 应用 152
- 4.5 PixelRatio 154
  - 4.5.1 PixelRatio 介绍 154
  - 4.5.2 PixelRatio 应用 155
- 4.6 AppStateIOS 156
  - 4.6.1 AppStateIOS 介绍 156
  - 4.6.2 AppStateIOS 实例 156
- 4.7 StatusBarIOS 157
  - 4.7.1 StatusBarIOS 介绍 157
  - 4.7.2 StatusBarIOS 应用 157
- 4.8 NetInfo 158
  - 4.8.1 NetInfo 介绍 159
- 4.8.2 NetInfo 示例 159
- 4.9 CameraRoll 159
  - 4.9.1 CameraRoll 介绍 160
  - 4.9.2 CameraRoll 应用 160
  - 4.9.3 react-native-camera 167
- 4.10 VibrationIOS 170
- 4.11 Geolocation 171
  - 4.11.1 Geolocation 介绍 171
  - 4.11.2 Geolocation 应用 172
- 4.12 数据请求 173
  - 4.12.1 XMLHttpRequest 174
  - 4.12.2 Fetch 174
- 4.13 定时器 175
  - 4.13.1 setTimeout 175
  - 4.13.2 setInterval 176
  - 4.13.3 setImmediate 176
  - 4.13.4 使用 requestAnimationFrame 开发进度条 176
  - 4.13.5 完整代码 177

第 5 章 Native 扩展 178

- 5.1 通信机制 178
  - 5.1.1 模块配置映射 178
  - 5.1.2 通信流程 180
- 5.2 自定义 Native API 组件 183
  - 5.2.1 模块和方法定义 184
  - 5.2.2 回调函数 185
  - 5.2.3 线程 187
  - 5.2.4 常量导出 188
  - 5.2.5 事件 189
  - 5.2.6 实战 190
- 5.3 构建 Native UI 组件 196
  - 5.3.1 概述 196
  - 5.3.2 UI 组件的定义 196
  - 5.3.3 UI 组件属性 197
  - 5.3.4 组件方法 199

        5.3.5　事件 199
        5.3.6　实例 201

第6章　组件封装 208
  6.1　二级菜单组件 208
      6.1.1　静态组件的实现 209
      6.1.2　实现组件的复用和封装 213
      6.1.3　应用二级菜单组件 219
  6.2　日历组件 220
      6.2.1　开发日历组件 221
      6.2.2　应用日历组件 226
  6.3　开源组件 227
      6.3.1　react-native-swiper 228
      6.3.2　react-native-modal 231

第三部分　App 更新和上架篇

第7章　热更新和上架 236
  7.1　动态更新 236
      7.1.1　初始化设置 236
      7.1.2　更新逻辑 238
      7.1.3　回滚策略 240
  7.2　App 上架 240
      7.2.1　证书生成 241
      7.2.2　注册应用 248
      7.2.3　上传应用 249

第四部分　实战篇

第8章　企业内部通讯录应用开发 254
  8.1　需求提出 254
  8.2　技术架构 255
  8.3　服务器端设计和开发 256
      8.3.1　服务器端整体设计 257
      8.3.2　用户数据模型设计 257
      8.3.3　公告数据模型设计 259
      8.3.4　服务路由设计 260

      8.3.5　创建项目 260
      8.3.6　app.js 入口文件 262
      8.3.7　加载服务模块到内存 263
      8.3.8　工具类开发 264
      8.3.9　用户信息接口 265
      8.3.10　公告消息接口 270
      8.3.11　建议 272
  8.4　客户端设计和开发 272
      8.4.1　客户端设计 272
      8.4.2　工具组件和服务 274
      8.4.3　添加依赖库 275
      8.4.4　程序入口和登录 277
      8.4.5　联系人列表 286
      8.4.6　公告功能 294
      8.4.7　管理功能 299
      8.4.8　关于 316
      8.4.9　建议 318

第9章　基于 LBS 的应用开发 319
  9.1　功能设计 319
      9.1.1　需求确定 319
      9.1.2　开发目录结构 320
  9.2　程序入口和工具模块 321
      9.2.1　注册应用程序 321
      9.2.2　工具模块 322
      9.2.3　Nearby 组件入口 323
  9.3　列表组件开发 327
      9.3.1　通用列表组件开发 327
      9.3.2　完成列表页 334
  9.4　详情页组件开发 336
  9.5　WebView 地图模块开发 338
  9.6　综合效果 341

第10章　豆搜 App 342
  10.1　豆瓣 API 342
      10.1.1　熟悉豆瓣 API 342

    10.1.2　图书、电影、音乐 API……… 343
10.2　应用设计…………………………… 345
    10.2.1　功能设计……………………… 345
    10.2.2　模块划分……………………… 345
10.3　公共模块开发……………………… 347
    10.3.1　工具类开发…………………… 347
    10.3.2　服务列表……………………… 348
    10.3.3　Navigator 封装……………… 349
    10.3.4　公共头封装…………………… 350

    10.3.5　WebView 封装……………… 352
    10.3.6　搜索框封装…………………… 353
10.4　功能开发…………………………… 354
    10.4.1　入口组件……………………… 354
    10.4.2　图书列表页开发……………… 356
    10.4.3　图书详情页开发……………… 361
    10.4.4　电影模块开发………………… 363
    10.4.5　音乐模块开发………………… 368
10.5　完成豆搜 App……………………… 372

第一部分
基础语法篇

- 第1章 React Native简介
- 第2章 React Native开发基础

第 1 章
React Native简介

React Native一经Facebook开源,就引起了业界的关注,越来越多的开发者开始尝试在生产中使用它。React Native为JavaScript开发跨终端应用提供了更加丰富的想象空间。下面就开始我们的React Native开发之旅吧。

1.1 环境搭建

工欲善其事,必先利其器。首先,我们需要搭建开发环境(整本书都是以Mac OS X系统为基础的)。React Native主要依赖的环境如下所示。

- ❑ Mac OS X操作系统。
- ❑ 推荐使用Xcode 6.4或者更高版本。
- ❑ 安装Node.js 4.0或者最新版本。
- ❑ 建议使用Homebrew安装:watchman和flow。

下面我们一步步来安装开发环境。

1.1.1 安装Node.js

打开浏览器,在浏览器中输入地址https://nodejs.org/,此时打开的是Node.js的官网,从中可以看到最新的版本以及下载按钮。

这里下载的是node-v4.1.0.pkg。等待下载完成后,我们开始安装Node.js。双击node-v4.1.0.pkg文件,将弹出如图1-1所示的界面。

图1-1　Node.js安装启动界面

在图1-1中，我们可以看到This package will install Node.js v4.1.0 and npm v2.14.3 into /usr/local/这句话，这表明将会安装Node.js 4.1.0版本和npm 2.14版本。

接着点击"继续"按钮，会看到许可界面，再点击"继续"按钮，将看到许可提示，如图1-2所示。

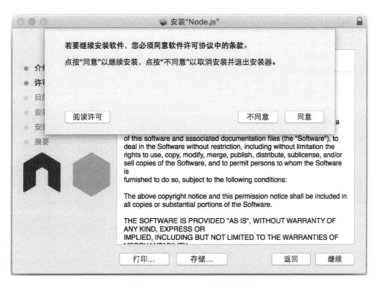

图1-2　Node.js安装许可

点击"同意"按钮，会出现如图1-3所示的界面。

图1-3　确认为用户安装Node.js

这里我们点击"继续"按钮，得到的界面如图1-4所示。

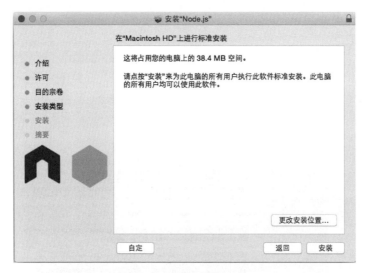

图1-4　选择安装目录

在图1-4中，我们不更改安装位置，直接使用默认值即可。然后点击"安装"按钮，输入你的电脑的密码（一般用户都设置了电脑密码），再点击"安装软件"开始安装程序。如图1-5所示，这里显示了"安装成功"，表示Node.js安装成功，此时直接点击"关闭"按钮即可。

图1-5　安装成功界面

在图1-5中,我们可以看到如下信息:

```
Node.js was installed at

  /usr/local/bin/node

npm was installed at

  /usr/local/bin/npm

Make sure that /usr/local/bin is in your $PATH.
```

以上信息表明,Node.js安装在/usr/local/bin/node目录下,npm安装在/usr/local/bin/npm目录下。我们可以通过Mac终端命令切换到/usr/local/bin/目录下,从中可以看到安装的Node.js和npm,如图1-6所示。上述信息同时告诉我们,要确保/usr/local/bin在我们的环境变量%PATH中,一般默认都在。

图1-6　Node.js和npm安装目录

此时，我们在终端中输入命令node -v，就可以看到刚才安装的Node.js的版本（即v4.1.0），表示Node.js安装成功，如图1-7所示。现在我们就可以在终端中写些简单的程序了。需要说明的是，这里我们就不介绍Node.js了，读者可自行查阅相关资料。

图1-7　Node.js

1.1.2　安装React Native

Node.js安装完成后，可以使用它干很多事情。但是，本书的主题是React Native，所以我们要继续搭建React Native的开发环境。

在正式安装React Native之前，需要确保以下环境是可用的。

- Node.js已经安装且在环境变量中。如果没有安装，可以参考1.1.1节。
- Xcode已经安装且版本最好是6.4以上版本。Xcode的安装这里就不做介绍了。
- 推荐安装Homebrew，同时通过Homebrew安装watchman和flow。

通过Homebrew安装watchman和flow的命令如下：

```
$ brew install watchman
$ brew install flow
```

现在万事具备，只欠东风。我们通过npm安装react-native-cli的命名行工具。在Mac终端中输入如下命令，其中-g表示全局安装：

```
$ npm install -g react-native-cli
```

如果在安装过程中发现需要管理员权限，可以给命令添加sudo，然后输入管理员密码即可：

```
$ sudo npm install -g react-native-cli
```

安装成功后，看到的界面如图1-8所示。

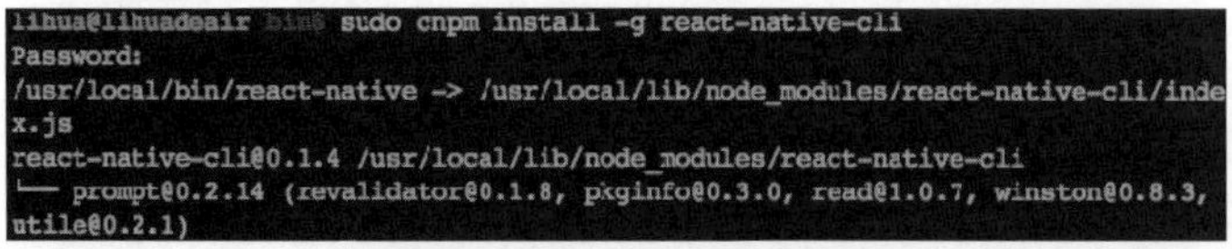

图1-8　安装react-native-cli

如果安装耗时较长，可以采用淘宝镜像安装，即如图1-8所示的cnpm。

1.1.3　使用 NVM 管理 Node.js 版本

因为需要经常切换Node.js版本，所以建议使用NVM（Node.js Version Manager）来管理Node.js版本。NVM在GitHub上的地址是https://github.com/cnpm/nvm（这里使用cnpm的NVM）。

首先，我们使用git命令将代码克隆下来。例如，在命令行中输入git clone命令：

```
$ git clone https://github.com/creationix/nvm
```

为了临时使用nvm命令（只针对当前bash），在终端中输入如下命令：

```
$ cd nvm（新版没有该目录）
$ source nvm.sh
```

这样我们就可以用nvm对Node.js和io.js进行版本管理了。在终端中输入nvm命令，可以看到命令帮助：

```
$ nvm
Node Version Manager
Usage:
  nvm help                        Show this message
  nvm --version                   Print out the latest released version of nvm
  nvm install [-s] <version>      Download and install a <version>, [-s] from source. Uses .
                                    nvmrc if available
  nvm uninstall <version>         Uninstall a version
  nvm use <version>               Modify PATH to use <version>. Uses .nvmrc if available
  nvm run <version> [<args>]      Run <version> with <args> as arguments. Uses .nvmrc if
                                    available for <version>
  nvm current                     Display currently activated version
  nvm ls                          List installed versions
  nvm ls <version>                List versions matching a given description
  nvm ls-remote                   List remote versions available for install
  nvm deactivate                  Undo effects of 'nvm' on current shell
  nvm alias [<pattern>]           Show all aliases beginning with <pattern>
  nvm alias <name> <version>      Set an alias named <name> pointing to <version>
  nvm unalias <name>              Deletes the alias named <name>
  nvm reinstall-packages <version>  Reinstall global 'npm' packages contained in <version> to
                                    current version
  nvm unload                      Unload 'nvm' from shell
  nvm which [<version>]           Display path to installed node version. Uses .nvmrc if
                                    available
Example:
  nvm install v0.10.32            Install a specific version number
  nvm use 0.10                    Use the latest available 0.10.x release
  nvm run 0.10.32 app.js          Run app.js using node v0.10.32
  nvm exec 0.10.32 node app.js    Run 'node app.js' with the PATH pointing to node v0.10.32
  nvm alias default 0.10.32       Set default node version on a shell
Note:
  to remove, delete, or uninstall nvm - just remove ~/.nvm, ~/.npm, and ~/.bower folders
```

我们使用nvm命令来查看远程Node.js版本有哪些：

```
$ nvm ls-remote
```

同时可以使用nvm ls命令查看当前本机已经安装的所有Node.js的版本，如图1-9所示。

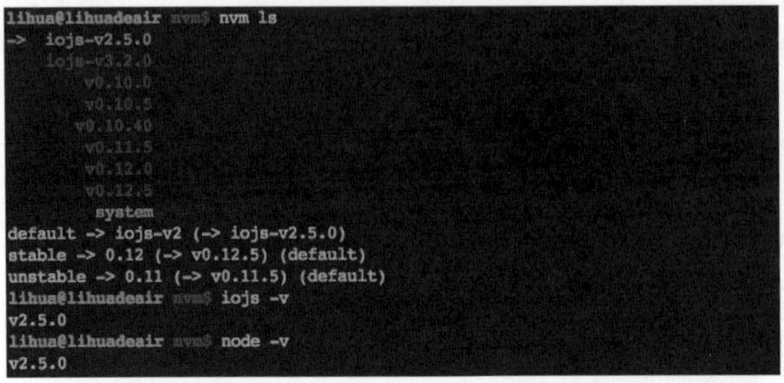

图1-9　查看当前本机已经安装的所有Node.js的版本

1.1.4　创建项目

搭建好React Native环境后，现在来创建项目Hello。在命令行中使用react-native init命令：

```
$ react-native init Hello
This will walk you through creating a new React Native project in /Users/lihua/work/Hello

> bufferutil@1.2.1 install /Users/lihua/work/Hello/node_modules/react-native/node_modules/ws/node_modules/bufferutil
> node-gyp rebuild

  CXX(target) Release/obj.target/bufferutil/src/bufferutil.o
  SOLINK_MODULE(target) Release/bufferutil.node

> 这里省略了安装过程的信息

To run your app on iOS:
   Open /Users/lihua/work/Hello/ios/Hello.xcodeproj in Xcode
   Hit Run button
To run your app on Android:
   Have an Android emulator running, or a device connected
   cd /Users/lihua/work/Hello
   react-native run-android
```

到目前为止，我们已经创建好了项目，可以看到Open /Users/xxx/Hello/ios/Hello.xcodeproj in Xcode，这提示我们使用Xcode打开项目。

进入刚才创建的Hello项目，点击Hello.xcodeproj打开该项目。需要注意的是，xcodeproj类型的文件是Xcode的项目文件。使用快捷键cmd（⌘）+R启动项目，这时可以看到启动了一个终端React Packager和一个模拟器。如果想了解React Packager，可以翻阅到2.3节查看。

项目启动完成后，就可以看到模拟器上显示了Welcome to React Native!，这说明项目运行成功，如图1-10所示。

图1-10　Hello World

我们使用Sublime Text打开index.ios.js文件，将Welcome to React Native!修改为"开始React Native编程之旅"，此时再使用cmd（⌘）+R快捷键刷新模拟器，会立即显示"开始React Native编程之旅"的字样。这说明，修改JavaScript文件起作用了。

1.2　从 React 到 React Native

React Native是基于React设计的，因此，了解React有助于我们开发React Native应用。

1.2.1　React 简介

React的GitHub地址是https://github.com/facebook/react。截至2015年7月26日下午3点，React已经获得25451个star、3733个fork，以及数百个issues和pull requests。这说明React受到了很大的重视，尤其是前端工程师。React的官方地址是：http://facebook.github.io/react/。

目前，React的最新版本是React v0.14.0，其官方的介绍是A JAVASCRIPT LIBRARY FOR BUILDING USER INTERFACES。可以看到，React提出的是一个新的开发模式和理念，它强调的是"用户界面"。因此，很多开发者将React同Angular、Ember、Backbone等前端框架进行比较是不太合理的。

毕竟，大家思考的出发点就不一致，React希望将功能分解化，让开发变得像搭积木一样，快速且可维护。还有开发者开了个玩笑，说React抛弃了View（视图）和Controller（控制器）的分离实践，一夜回到了"解放前"。这说明，有很多开发者对React还是存在一定疑虑和思考的。

接受一个新事物需要一定的时间，同样接受一个新的开发模式，并且该开发模式丢弃了一些既有的经验，需要一定的勇气和踩坑的精神。

React主要有如下3个特点。

- 作为UI（Just the UI）：React可以只作为视图（View）在MVC中使用。并且在已有项目中，很容易使用React开发新功能。
- 虚拟DOM（Virtual DOM）：这是React的一个亮点，可以很好地优化视图的渲染和刷新。当然，它也可以在Node.js服务器端和React Native中使用。虚拟DOM是React最重要的一个特性。以前，我们更新视图时，需要先清空DOM容器中的内容，然后将最新的DOM和数据追加到容器中，现在React将这一操作放进了内存。React认为内存的操作远比视图全部更新来得高效。随着浏览器的迭代，事实情况更是如此，内存相比视图刷新要廉价得多。React将视图变化放进内存进行比较（就是虚拟DOM的比较），计算出最小更新的视图，然后将该差异部分进行更新以完成整个组件的渲染。这就是React如此高效的原因。
- 数据流（Data Flow）：React实现了单向的数据流，并且相对于传统的数据绑定而言，React更加灵活、便捷。

现在我们对React已经有了初步的了解，那么，学习React需要掌握哪些知识呢？

- JSX语法知识：JSX的官方解释是其语法类似于XML（an XML-like syntax），这也是前端工程师更容易理解JSX的原因。因为HTML是XML的子集，前端工程师对HTML更为熟悉。
- ES6相关知识：因为ES6增加了很多语法特性和新功能，可以使用ES6更加快速地进行功能开发。
- 前端基础知识：当然，最为基础的是需要具备基本的前端知识，其中CSS以及JavaScript比较重要。

现在，我们举3个简单的例子来说明React的用法。关于IDE，可以使用WebStorm，因为WebStorm拥有比较丰富的插件。

1. 简单组件和数据传递

我们可以使用this.props来做简单的数据传递，具体代码如下：

```
<!DOCTYPE html>
<html>
<head lang="en">
  <meta charset="UTF-8">
  <title>React第一个例子</title>
  <script type="text/javascript" src="react.js"></script>
  <script type="text/javascript" src="JSXTransformer.js"></script>
</head>
<body>
  <div id="example"></div>
  <script type="text/jsx">
    var HelloMessage = React.createClass({
      render: function() {
```

```
        return <h1>Hello {this.props.name}!</h1>;
      }
    });
    React.render(<HelloMessage name="React" />, document.getElementById('example'));
    </script>
  </body>
</html>
```

我们从http://facebook.github.io/react/downloads.html上下载react.js和JSXTransformer.js这两个文件，其中react.js是React的核心文件，JSXTransformer.js是将JSX转译成JavaScript和HTML的工具。在上面的例子中，我们使用的是JSXTransformer.js直接引入，这样的话，就可以在运行时解析转换JSX。当然，在生产环境中，我们可以将JSXTransformer.js所做的工作转到线下，做个预编译。例如，可以使用Node.js做预编译，可以安装react-tools工具：

```
$ npm install -g react-tools
```

这样，代码经过了预编译、压缩和合并后，会提高网络的加载速度，减少流量宽带的浪费，优化了用户体验。

现在，我们分析上面的代码，它主要做了两件事。

- 定义了一个组件HelloMessage。HelloMessage可以传入name属性，同可以将内容用h1标签渲染。
- 使用React.render方法将组件渲染到id为example的div内。render方法有两个参数，第一个参数就是要渲染的组件内容，第二个就是要渲染到的目标节点。

此时打开浏览器，可以看到效果：Hello React!。我们打开Chrome console工具，看到渲染后的文档内容如图1-11所示。

```
▼<html>
  ▶<head lang="en">...</head>
  ▼<body>
    ▼<div id="example">
      ▼<h1 data-reactid=".0">
          <span data-reactid=".0.0">Hello </span>
          <span data-reactid=".0.1">React</span>
          <span data-reactid=".0.2">!</span>
        </h1>
      </div>
    ▶<script type="text/jsx">...</script>
  </body>
</html>
```

图1-11　React渲染DOM

可以看到，React将文本使用span包裹，然后嵌入了id为example的div。

2. 通过this.state更新视图

通过改变this.state来更新视图，具体代码如下：

```
<script type="text/jsx">
  var Timer = React.createClass({
    /*初始状态*/
    getInitialState: function() {
      return {secondsElapsed: 0};
    },
```

```
    tick: function() {
      this.setState({secondsElapsed: this.state.secondsElapsed + 1});
    },
    /*组件完成装载*/
    componentDidMount: function() {
      this.interval = setInterval(this.tick, 1000);
    },
    /*组件将被卸载*/
    componentWillUnmount: function() {
      clearInterval(this.interval);
    },
    render: function() {
      return (
        <div>Seconds Elapsed: {this.state.secondsElapsed}</div>
      );
    }
  });
  React.render(<Timer />, document.getElementById('example'));
</script>
```

这里我们省略了HTML的内容，可以参考第一个例子。上面的代码只做了一件事：更改组件的状态this.state。

在浏览器中运行效果，可以发现每隔一秒，secondsElapsed增加1。现在，我们来看看上面的代码。我们通过React.createClass创建了一个组件。

其实，这里已经涉及一个组件的生命周期了。getInitialState是组件的初始状态，必须返回一个对象或者null对象。在getInitialState中，我们可以准备组件需要的数据以及后期需要更新的数据模型，也就是说getInitialState返回的对象是挂载在this.state上的。render方法的作用是渲染视图。这里render使用的数据是this.state，这样我们可以通过更新this.state来更新视图。componentDidMount是组件加载完成的状态，这里我们可以改变组件的状态（this.state）。

在上面的代码中，componentDidMount设置了interval，每隔一秒钟，secondsElapsed加1。componentWillUnmount在组件将被卸载时调用。我们可以在componentWillUnmount里清除定时器this.interval。

3. 简单应用

示例代码如下：

```
var ShowEditor = React.createClass({
  getInitialState: function() {
    return {value: '你可以在这里输入文字'};
  },
  handleChange: function() {
    this.setState({value: React.findDOMNode(this.refs.textarea).value});
  },
  render: function() {
    return (
```

```
      <div>
        <h3>输入</h3>
        <textarea style={{width:300, height:150, outline:'none'}}
          onChange={this.handleChange}
          ref="textarea"
          defaultValue={this.state.value} />
        <h3>输出</h3>
        <div>
          {this.state.value}
        </div>
      </div>
    );
  }
});
React.render(<ShowEditor />, document.getElementById('example'));
```

上面的代码在浏览器中的运行效果如图1-12所示。

图1-12 React演示效果

现在我们来分析下上面的代码。ShowEditor做了一件事：用户在textarea输入文字的同时，会在div中及时输出textarea中的文字。

这里有4个地方需要我们关注下。

❑ 可以通过state来修改视图的状态从而改变视图。
❑ textarea上绑定了onChange的事件监听，其目的是通过setState改变this.state.value。
❑ textarea添加了ref属性，这样我们就可以通过this.refs.textarea引用textarea对象了。
❑ React提供findDOMNode的方法，通过它可以找到React的DOM。

通过上面这3个简单的例子，我们对React有了初步的认识。

1.2.2 React Native 简介

前面我们介绍了React的一些特性，展示了几个简单的例子。其实React与React Native具有非常密切的关系，现在就来介绍下React在Native上的应用——React Native。React Native在GitHub

上地址是：https://github.com/facebook/react-native。

截至2015年11月24日下午1点，React Native已经获得了22839个star和3644个fork，这说明React Native十分受大家关注。目前（2015年11月），React Native的最新版本是v0.15.0，我们可以到https://github.com/facebook/react-native/releases关注最新版本的发布情况。

React Native的官网地址是https://facebook.github.io/react-native/，其官网的介绍是使用React构建Native应用的框架（A FRAMEWORK FOR BUILDING NATIVE APPS USING REACT）。

这说明，React Native采用的语法也是React。React Native的目标是高效跨平台地开发Native应用，同时，也强调了"一次学习，多个平台编写代码"。也就是说，React Native不是"一次编码，多处运行"。如图1-13所示，我们可以清楚地看到React Native是构建在React和JSX基础上的。因此，只要基本掌握了React和JSX，同时补充相关平台（iOS、Android、Web）的知识，就能开发Native应用和Web应用。

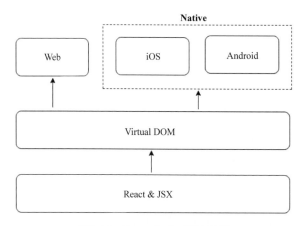

图1-13　React Native设计思路

现在，我们来看一个简单的例子。在1.1.2节中，我们已经知道如何创建一个项目，这里创建一个名为demo的React Native项目。

```
$ react-native init demo
```

然后使用Xcode打开该项目，按照如下代码修改index.ios.js文件：

```
'use strict';

var React = require('react-native');
var {
  AppRegistry,
  StyleSheet,
  Text,
  View,
  Image,
```

```
} = React;

var demo = React.createClass({
  render: function() {
    return (
      <View style={styles.container}>
        <Text style={styles.welcome}>欢迎来到React Native的世界</Text>
        <Image style={{width:50,height:50, resizeMode: Image.resizeMode.contain}}
          source={{uri:'https://facebook.github.io/react-native/img/header_logo.png'}}>
        </Image>
      </View>
    );
  }
});

var styles = StyleSheet.create({
  container: {
    flex: 1,
    justifyContent: 'center',
    alignItems: 'center',
    backgroundColor: '#05A5D1',
  },
  welcome: {
    fontSize: 20,
    color:'#fff'
  },
});

AppRegistry.registerComponent('demo', () => demo);
```

接着在Xcode中启动项目（快捷键：cmd（⌘）+R），运行效果如图1-14所示。

图1-14　React Native演示效果

下面我们来分析上面的代码。使用require引入react-native模块，然后定义了AppRegistry、StyleSheet、Text、View和Image这5个对象。这5个对象其实是React Native的3个组件（Text、View和Image）和2个API（AppRegistry和StyleSheet）。如果想了解更多的组件和API，可以翻阅到第3章和第4章。

其实，上面的var形式的对象也可以这样定义：

```
//var {
//   AppRegistry,
//   StyleSheet,
//   Text,
//   View,
//   Image,
//} = React;

var AppRegistry = React.AppRegistry;
var StyleSheet = React.StyleSheet;
var Text = React.Text;
var View = React.View;
var Image = React.Image;
```

可以看出，使用var {} = React的形式更为简洁。

然后使用React.createClass创建了一个组件，其中render是视图渲染的方法，渲染的内容是一行文本"欢迎来到React Native的世界"和一个图片（React Native Logo）。StyleSheet.create创建的是一个样式表的类，里面包含container和welcome这两个样式对象。同时，我们也看到组件上的style属性包含了一个对象，即style={对象}的形式。这里需要说明的是，样式书写与CSS不一样，我们需要使用驼峰命名规则。关于React Native的样式、布局及语法，我们将在第2章中详细介绍。

图1-15展示了React Native的开发流程。首先，我们需要根据业务来划分组件。这时候，组件的颗粒度需要根据业务分工和开发难度来划分，否则，后期的维护成本就会相对较高。我们希望，根据不同的功能开发不同的入口组件，这样的话，每一个功能或者功能集合就能独立出来，功能的移植性好。关于组件的功能开发，我们将在第3章和第6章中详细介绍，第8章和第9章都会以实战的形式展示。最后，为了更新和发布应用，我们需要设计和开发一套符合业务需求的打包更新机制。关于打包和热更新，我们将在第7章中介绍。

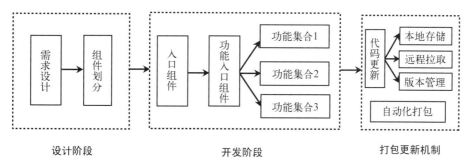

图1-15 React Native开发流程

1.3 为什么要使用 React Native

首先，我们来回答为什么会出现React Native？这是因为移动设备的环境要比Web环境复杂得多，也就导致了Native开发的成本要高。随着前端的发展，已经拥有了大量的前端从业人员，但是移动端开发人员相对较少。一些公司为了寻求App开发效率、成本、体验之间的平衡，从而选择了Hybird App的开发方案。这样做的一个好处就是，既能拥有高效的开发效率和较多的开发人员，又能快速更新App。但是，这只是一种平衡方案。更多的时候，在资源较为丰富的情况下，更倾向于把体验做好。然而，在WebView中嵌入HTML页面存在一些性能和体验上的弱势。这也为技术发展提出了一个新的挑战：如何将开发成本和用户体验做到更好的平衡？

因此，Facebook提出了React Native的解决方案。也正是因为React Native的跨平台解决的特性和使用JavaScript作为开发语言而赢得了众多开发者的关注。很多时候，前端都有一种乐观的想法：HTML5可以替代原生应用。但是，实际上，HTML5应用在用户体验和性能上比原生应用弱一些。这就是React Native的切入点。React Native不仅可以使用前端开发的模式来开发应用，还能调用原生应用的UI组件和API。所以说，React Native兼顾了开发效率，提高了用户体验。这也为前端开发者进入原生开发领域降低了门槛。

1.4 如何学习 React Native

目前，React Native更新的速度较快，文档方面还不是很全。如果开发一款小型的App，掌握React Native的组件和API就已经足够了。如果学习和实践中遇到问题，可以到 React Native GitHub issues上搜索，其中有很多解决方法。下面是关于React Native的4个比较重要的地址。

- ❑ React Native官方网站：http://facebook.github.io/react-native/。
- ❑ React Native版本发布：https://github.com/facebook/react-native/releases。
- ❑ React Native GitHub地址：https://github.com/facebook/react-native。
- ❑ 疑难问题搜索：https://github.com/facebook/react-native/issues。

1.5 说明

因为React Native Android版本还不稳定，本书所有实例都是基于iOS开发的。Android环境搭建可以参考：http://facebook.github.io/react-native/docs/android-setup.html。此外，因为React Native是跨平台的，在iOS和Android平台上只是一些组件和API存在差异，所以学习了React Native iOS开发后，开发React Native Android也是比较容易的事。但是，需要我们针对iOS和Android平台学习一些基础知识。

第 2 章
React Native开发基础

经过上一章的学习，大家对React Native都有了一定的认识。光认识还不够，我们还需为React Native应用的开发打下扎实的基础。本章中，我们主要介绍React Naitve应用开发中用到的最基础的内容，包括处理页面布局的flexbox、处理页面结构的JSX以及React Native的开发向导。

2.1 flexbox 布局

flexbox是React Native应用开发中必不可少的内容，也是最常用的内容。下面我们将对flexbox的基础知识以及在React Native应用中的实战进行详细的介绍。

2.1.1 介绍

做过Web开发的人员都清楚，传统的页面布局基于盒子模型，依赖定位属性、流动属性和显示属性来解决。对于一些伸缩性的布局，处理起来很是麻烦，于是在2009年，由W3C组织提出来一种新的布局方案，即flexbox布局。该布局可以简单快速地完成各种伸缩性的设计。那么，flexbox是什么呢？flexbox是Flexible Box的缩写，即为弹性盒子布局，可以为传统的盒子模型布局带来更大的灵活性。关于该布局，W3C的最近更新时间是2015年5月14日。在http://caniuse.com/中显示，该布局对浏览器的支持情况如图2-1所示，从中可以看出，目前主流的浏览器都已支持它。如果只考虑移动端开发的同学们，就可以随意使用它了。

图2-1 浏览器对flexbox布局的支持情况

说明　图中红色表示不支持，黄色表示支持部分功能，绿色表示完全支持，右上角带有"–"标记的，表示需要加浏览器前缀。

由于本书为黑白印刷，所以这里详细说明下哪些版本归属于哪种颜色。

- 红色：IE8和IE9。
- 黄色：IE10、Android Browser 4.1、Android Browser 4.3。
- 绿色：除了上面的红色和黄色，剩下的都为绿色。

2.1.2　布局模型

flexbox布局由伸缩容器和伸缩项目组成。任何一个元素都可以指定为flexbox布局，其中设为display:flex或display:inline-flex的元素称为伸缩容器。伸缩容器的子元素称为伸缩项目，伸缩项目使用伸缩布局模型来排版。伸缩布局模型与传统的布局不一样，它按照伸缩流的方向布局。为了更好地了解伸缩流布局，我们先看一张来自W3C的图片，如图2-2所示。

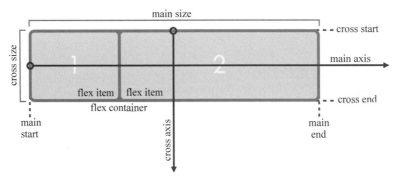

图2-2　伸缩布局图

默认情况下，伸缩容器由两根轴组成，即主轴（main axis）和交叉轴（cross axis），其中主轴的开始位置叫做main start，结束位置叫main end。交叉轴的开始位置叫cross start，结束位置叫cross end。伸缩项目在主轴上占据的空间叫main size，在交叉轴上占据的空间叫cross size。根据设置情况的不同，主轴既可以是水平轴，也可以是垂直轴。不论哪个轴作为主轴，默认情况下，伸缩项目总是沿着主轴，从主轴开始位置到主轴结束位置进行排列。flexbox目前还处于草稿状态，所以在使用flexbox布局的时候，需要加上各个浏览器的私有前缀，即-webkit、-moz、-ms、-o等。本章中为了书写方便，一律去掉了前缀，大家在使用时切记不要忘了加前缀。

2.1.3　伸缩容器属性

伸缩容器支持的属性有：

- display
- flex-direction
- flex-wrap
- flex-flow
- justify-content
- align-items
- align-content

下面简要介绍这几个属性。

1. display

该属性用来指定元素是否为伸缩容器，其语法为：

display:flex | inline-flex

HTML代码为：

```
<span class="flex-container"></span>
```

下面简要介绍这两个属性值的含义。

- **flex**：这个值用于产生块级伸缩容器，示例CSS代码如下：

    ```
    .flex-container {
      display:flex;
    }
    ```

- **inline-flex**：这个值用于产生行内级伸缩容器，示例CSS代码如下：

    ```
    .flex-container {
      display: inline-flex;
    }
    ```

说明　这个时候，CSS的columns在伸缩容器上没有效果，float、clear和vertical-align在伸缩项目上没有效果。

2. flex-direction

该属性用于指定主轴的方向，其语法是：

flex-direction: row | row-reverse | column | column-reverse

HTML代码如下：

```
<span class="flex-container">
  <span class="flex-item">1</span>
  <span class="flex-item">2</span>
```

```
      <span class="flex-item">3</span>
</span>
```

下面简要介绍这4个属性值的含义。

- **row**（默认值）：伸缩容器若为水平方向轴，伸缩项目的排版方式为从左向右排列。示例CSS代码如下：

```
.flex-container {
  display: flex;
  flex-direction: row;
}
```

效果如图2-3所示。

图2-3　flex-direction为row的效果图

> **说明**　flex-directon的默认值为row，所以伸缩容器设置好后，如果是横向伸缩，则无需定义flex-direction。

- **row-reverse**：伸缩容器若为水平方向轴，伸缩项目的排版方式为从右向左排列。示例CSS代码如下：

```
.flex-container {
  display: flex;
  flex-direction: row-reverse;
}
```

效果如图2-4所示。

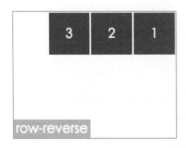

图2-4　flex-direction为reverse的效果图

- **column**：伸缩容器若为垂直方向轴，伸缩项目的排版方式为从上向下排列。示例CSS代码如下：

```
.flex-container {
  display: flex;
  flex-direction: column;
}
```

效果如图2-5所示。

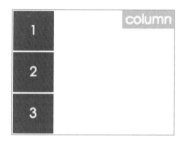

图2-5　flex-direction为column的效果图

- **column-reverse**：伸缩容器若为垂直方向轴，伸缩项目的排版方式为从下向上排列。示例CSS代码如下：

```
.flex-container {
  display: flex;
  flex-direction: column-reverse;
}
```

效果如图2-6所示。

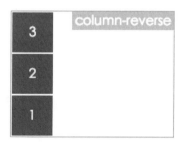

图2-6　flex-direction为column-reverse的效果图

3. flex-wrap

该属性主要用来指定伸缩容器的主轴线方向空间不足的情况下，是否换行以及该如何换行，其语法为：

```
flex-wrap: nowrap | wrap | wrap-reverse
```

HTML代码如下：

```
<span class="flex-container">
  <span class="flex-item">1</span>
  <span class="flex-item">2</span>
  <span class="flex-item">3</span>
  <span class="flex-item">4</span>
  <span class="flex-item">5</span>
</span>
```

下面简要介绍这3个属性值的含义。

- **nowrap（默认值）**：即使空间不足，伸缩容器也不允许换行。示例CSS代码如下：

```
.flex-container {
  display: flex;
  flex-direction: row;
  flex-wrap: nowrap;
  width: 200px;
  height: 150px;
}
.flex-item {
  width: 50px;
  height: 50px;
}
```

效果如图2-7所示。

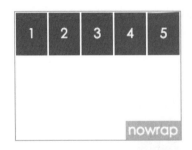

图2-7　flex-wrap为nowrap的效果图

- **wrap**：伸缩容器在空间不足的情况下允许换行。若主轴为水平轴，则换行的方向为从上到下。示例CSS代码如下：

```
.flex-container {
  display: flex;
  flex-direction: row;
  flex-wrap: wrap;
  width: 200px;
  height: 150px;
}
.flex-item {
  width: 50px;
```

```
      height: 50px;
}
```

效果如图2-8所示。

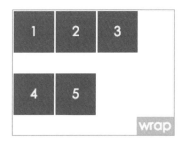

图2-8　flex-wrap为wrap的效果图

❑ **wrap-reverse**：伸缩容器在空间不足的情况下允许换行，若主轴为水平轴，则换行的方向为从下到上（和wrap相反）。示例CSS代码如下：

```
.flex-container {
    display: flex;
    flex-direction: row;
    flex-wrap: wrap-reverse;
    width: 200px;
    height: 150px;
}
.flex-item {
    width: 50px;
    height: 50px;
}
```

效果如图2-9所示。

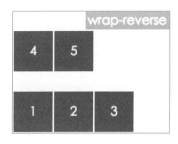

图2-9　flex-wrap为wrap-reverse的效果图

4. flex-flow

该属性是flex-direction和flex-wrap属性的缩写版本，它同时定义了伸缩容器的主轴和侧轴，其默认值为row nowrap。

该属性的语法为：

```
flex-flow: flex-direction flex-wrap
```

HTML代码如下：

```html
<span class="flex-container">
  <span class="flex-item">1</span>
  <span class="flex-item">2</span>
  <span class="flex-item">3</span>
  <span class="flex-item">4</span>
  <span class="flex-item">5</span>
</span>
```

下面我们以flex-flow:row wrap-reverse为例进行介绍，示例CSS代码如下所示：

```css
.flex-container {
  display: flex;
  flex-flow: row wrap-reverse;
  width: 200px;
  height: 150px;
}
.flex-item {
  width: 50px;
  height: 50px;
}
```

效果如图2-10所示。

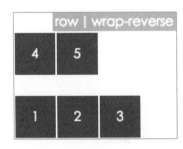

图2-10　flex-flow为row wrap-reverse的效果图

说明　图2-10和图2-9的效果一样。

5. justify-content

该属性用来定义伸缩项目沿主轴线的对齐方式，其语法为：

```
justify-content: flex-start | flex-end | center | space-between | space-around
```

HTML代码如下：

```
<span class="flex-container">
  <span class="flex-item">1</span>
  <span class="flex-item">2</span>
  <span class="flex-item">3</span>
</span>
```

下面简要介绍这5个属性值的含义。

❑ **flex-start**（默认值）：伸缩项目向主轴线的起始位置靠齐。示例CSS代码如下所示：

```
.flex-container {
  display: flex;
  flex-direction: row;
  justify-content: flex-start;
  width: 200px;
  height: 150px;
}
.flex-item {
  width: 50px;
  height: 50px;
}
```

效果如图2-11所示。

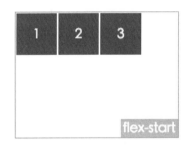

图2-11　justify-content为flex-start的效果图

❑ **flex-end**：伸缩项目向主轴线的结束位置靠齐。示例CSS代码如下：

```
.flex-container {
  display: flex;
  flex-direction: row;
  justify-content: flex-end;
  width: 200px;
  height: 150px;
}
.flex-item {
  width: 50px;
  height: 50px;
}
```

效果如图2-12所示。

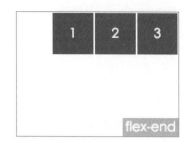

图2-12　justify-content为flex-end的效果图

❏ **center**：伸缩项目向主轴线的中间位置靠齐。示例CSS代码如下所示：

```css
.flex-container {
  display: flex;
  flex-direction: row;
  justify-content: center;
  width: 200px;
  height: 150px;
}
.flex-item {
  width: 50px;
  height: 50px;
}
```

效果如图2-13所示。

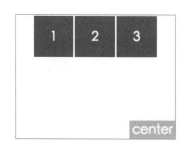

图2-13　justify-content为center的效果图

❏ **space-between**：伸缩项目会平均地分布在主轴线里。第一个伸缩项目在主轴线的开始位置，最后一个伸缩项目在主轴线的终点位置。示例CSS代码如下所示：

```css
.flex-container {
  display: flex;
  flex-direction: row;
  justify-content: space-between;
  width: 200px;
  height: 150px;
```

```
}
.flex-item {
  width: 50px;
  height: 50px;
}
```

效果如图2-14所示。

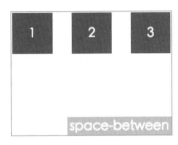

图2-14 justify-content为space-between的效果图

- **space-around**：伸缩项目会平均地分布在主轴线里，两端保留一半的空间。示例CSS代码如下：

```
.flex-container {
  display: flex;
  flex-direction: row;
  justify-content: space-around;
  width: 200px;
  height: 150px;
}
.flex-item {
  width: 50px;
  height: 50px;
}
```

效果如图2-15所示。

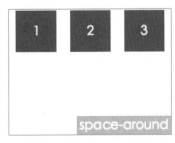

图2-15 justify-content为space-around的效果图

6. align-items

该属性用来定义伸缩项目在伸缩容器的交叉轴上的对齐方式，其语法为：

align-items: flex-start | flex-end | center | baseline | stretch

HTML代码如下:

```
<span class="flex-container">
  <span class="flex-item"  id="item1">1</span>
  <span class="flex-item"  id="item2">2</span>
  <span class="flex-item"  id="item3">3</span>
</span>
```

下面简要介绍这5个属性值的含义。

- **flex-start（默认值）**：伸缩项目向交叉轴的起始位置靠齐。示例CSS代码如下:

```
.flex-container {
  display: flex;
  flex-direction: row;
  align-items: flex-start;
  width: 200px;
  height: 150px;
}
.flex-item {
  width: 50px;
  height: 50px;
}
```

效果如图2-16所示。

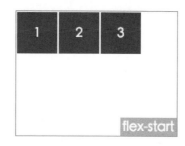

图2-16　align-items为flex-start的效果图

- **flex-end**：伸缩项目向交叉轴的结束位置靠齐。示例CSS代码如下:

```
.flex-container {
  display: flex;
  flex-direction: row;
  align-items: flex-end;
  width: 200px;
  height: 150px;
}
.flex-item {
  width: 50px;
  height: 50px;
}
```

效果如图2-17所示。

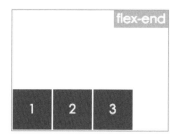

图2-17 align-items为flex-end的效果图

❑ **center**：伸缩项目向交叉轴的中间位置靠齐。示例CSS代码如下：

```
.flex-container {
  display: flex;
  flex-direction: row;
  align-items: center;
  width: 200px;
  height: 150px;
}
.flex-item {
  width: 50px;
  height: 50px;
}
```

效果如图2-18所示。

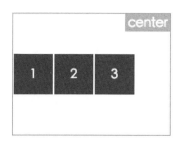

图2-18 align-items为center的效果图

❑ **baseline**：伸缩项目根据它们的基线对齐。示例CSS代码如下：

```
.flex-container {
  display: flex;
  flex-direction: row;
  align-items: baseline;
  width: 200px;
  height: 150px;
}
.flex-item {
  width: 50px;
```

```
  height: 50px;
}
#item1 {
  padding-top: 15px;
}
#item2 {
  padding-top: 10px;
}
#item3 {
  padding-top: 5px;
}
```

效果如图2-19所示。

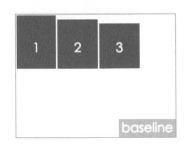

图2-19 align-items为baseline的效果图

❑ **stretch（默认值）**：伸缩项目在交叉轴方向拉伸填充整个伸缩容器。示例CSS代码如下：

```
.flex-container {
  display: flex;
  flex-direction: row;
  align-items: stretch;
  width: 200px;
  height: 150px;
}
.flex-item {
width: 50px;
}
```

效果如图2-20所示。

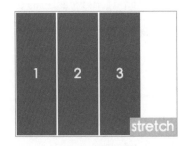

图2-20 align-items为stretch的效果图

说明　要看到该属性的效果，伸缩项目是不能设置高度的。

7. align-content

这个属性主要用来调整伸缩项目出现换行后在交叉轴上的对齐方式，类似于伸缩项目在主轴上使用justify-content，其语法为：

align-content: flex-start | flex-end | center | space-between | space-around | stretch

HTML代码如下：

```
<span class="flex-container">
  <span class="flex-item">1</span>
  <span class="flex-item">2</span>
  <span class="flex-item">3</span>
  <span class="flex-item">4</span>
  <span class="flex-item">5</span>
</span>
```

说明　flex-wrap: wrap这个一定要开启，且它在出现换行的情况下才能看到效果。下面提到的伸缩项目均指伸缩项目所在的行，因为这里调整的其实就是伸缩项目换行后每行之间的对齐方式。

下面简要介绍这6个属性值的含义。

❑ **flex-start**（默认值）：伸缩项目向交叉轴的起始位置靠齐。示例CSS代码如下：

```
.flex-container {
  display: flex;
  flex-direction: row;
  flex-wrap: wrap;
  align-content: flex-start;
  width: 200px;
  height: 150px;
}
.flex-item {
  width: 50px;
  height: 50px;
}
```

效果如图2-21所示。

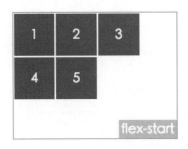

图2-21　align-content为flex-start的效果图

- **flex-end**：伸缩项目向交叉轴的结束位置靠齐。示例CSS代码如下：

```
.flex-container {
  display: flex;
  flex-direction: row;
  flex-wrap: wrap;
  align-content: flex-end;
  width: 200px;
  height: 150px;
}
.flex-item {
  width: 50px;
  height: 50px;
}
```

效果如图2-22所示。

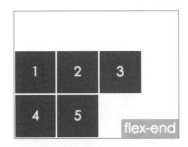

图2-22　align-content为flex-end的效果图

- **center**：伸缩项目向交叉轴的中间位置靠齐。示例CSS代码如下：

```
.flex-container {
  display: flex;
  flex-direction: row;
  flex-wrap: wrap;
  align-content: center;
  width: 200px;
  height: 150px;
}
.flex-item {
```

```
    width: 50px;
    height: 50px;
}
```

效果如图2-23所示。

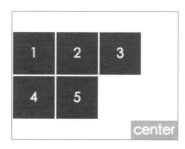

图2-23　align-content为center的效果图

❑ **space-between**：伸缩项目在交叉轴中平均分布。示例CSS代码如下：

```
.flex-container {
  display: flex;
  flex-direction: row;
  flex-wrap: wrap;
  align-content: space-between;
  width: 200px;
  height: 150px;
}
.flex-item {
  width: 50px;
  height: 50px;
}
```

效果如图2-24所示。

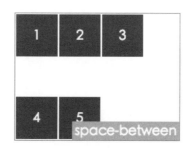

图2-24　align-content为space-between的效果图

❑ **space-around**：伸缩项目在交叉轴中平均分布，且在两边各有一半的空间。示例CSS代码如下：

```
.flex-container {
  display: flex;
  flex-direction: row;
```

```
  flex-wrap: wrap;
  align-content: space-around;
  width: 200px;
  height: 150px;
}
.flex-item {
  width: 50px;
  height: 50px;
}
```

效果如图2-25所示。

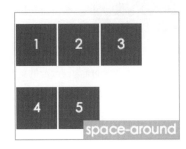

图2-25　align-content为space-around的效果图

- **stretch（默认值）**：伸缩项目将会在交叉轴上伸展以占用剩余的空间。示例CSS代码如下：

```
.flex-container {
  display: flex;
  flex-direction: row;
  flex-wrap: wrap;
  align-content: stretch;
  width: 200px;
  height: 150px;
}
.flex-item {
  width: 50px;
  height: 50px;
}
```

效果如图2-26所示。

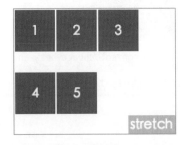

图2-26　align-content为stretch的效果图

2.1.4 伸缩项目属性

伸缩项目支持的属性有：

- order
- flex-grow
- flex-shrink
- flex-basis
- flex
- align-self

下面简要介绍这6个属性。

1. order

这个属性用于定义项目的排列顺序。数值越小，排列越靠前，默认值为0，其语法为：

order:integer

HTML代码如下：

```
<span class="flex-container">
  <span class="flex-item item1" >1</span>
  <span class="flex-item item2" >2</span>
  <span class="flex-item item3" >3</span>
  <span class="flex-item item4" >4</span>
  <span class="flex-item item5" >5</span>
</span>
```

示例CSS代码如下：

```
.flex-container {
  display: flex;
  flex-direction: row;
  flex-wrap: wrap;
  width: 200px;
  height: 150px;
}
.flex-item {
  width: 50px;
  height: 50px;
}
.item5 {
  order: -1;
}
```

效果如图2-27所示。

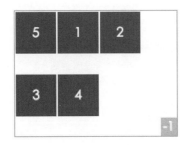

图2-27　order为–1的效果图

说明　示例中由于类名为item5的元素设置order属性为–1，所以排在第一位。

2．flex-grow

该属性定义伸缩项目的放大比例，默认值为0，即如果存在剩余空间，也不放大。如果所有伸缩项目的flex-grow设置为1，那么每个伸缩项目将设置为一个大小相等的剩余空间。如果你将其中一个伸缩项的flex-grow值设置为2，那么这个伸缩项目所占的剩余空间是其他伸缩项目所占剩余空间的两倍。其语法为：

```
flex-grow: number /* 其默认值为 0 */
```

HTML代码如下：

```html
<span class="flex-container">
  <span class="flex-item item1" >1</span>
  <span class="flex-item item2" >2</span>
  <span class="flex-item item3" >3</span>
  <span class="flex-item item4" >4</span>
  <span class="flex-item item5" >5</span>
</span>
```

示例CSS代码如下：

```css
.flex-container {
  display: flex;
  flex-direction: row;
  flex-wrap: wrap;
  width: 200px;
  height: 150px;
}
.flex-item {
  width: 50px;
  height: 50px;
}
.item5 {
  flex-grow:1;
}
```

效果如图2-28所示。

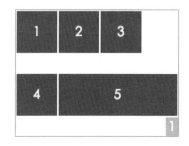

图2-28　flex-grow为1的效果图

3. flex-shrink

该属性用来定义伸缩项目的收缩能力，其语法如下：

```
flex-shrink: number /*其默认值为1 */
```

HTML代码如下：

```
<span class="flex-container">
  <span class="flex-item item1" >1</span>
  <span class="flex-item item2" >2</span>
  <span class="flex-item item3" >3</span>
  <span class="flex-item item4" >4</span>
  <span class="flex-item item5" >5</span>
</span>
```

示例CSS代码如下：

```
.flex-container {
  display: flex;
  flex-direction: row;
  flex-wrap: nowrap;
  width: 200px;
  height: 150px;
}
.flex-item {
  width: 50px;
  height: 50px;
}
.item5 {
  flex-shrink: 3;
}
```

效果如图2-29所示。

图2-29　flex-shrink为3的效果图

说明　本例子中，类名为item5的元素在空间不足的情况下，缩小为其他元素缩小大小的1/3。

---

4. flex-basis

该属性用来设置伸缩项目的基准值，剩余的空间按比率进行伸缩，其语法如下：

flex-basis: length | auto

其默认值为auto。

HTML代码如下：

```
<span class="flex-container">
  <span class="flex-item item1" >1</span>
  <span class="flex-item item2" >2</span>
  <span class="flex-item item3" >3</span>
  <span class="flex-item item4" >4</span>
  <span class="flex-item item5" >5</span>
</span>
```

示例CSS代码如下：

```
.flex-container {
  display: flex;
  flex-direction: row;
  flex-wrap: wrap;
  width: 200px;
  height: 150px;
}
.flex-item {
  width: 50px;
  height: 50px;
}
.item5 {
  flex-basis: 100px;
}
```

效果如图2-30所示。

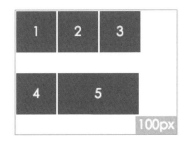

图2-30　flex-basis为100px的效果图

5. flex

该属性是flex-grow、flex-shrink和flex-basis这3个属性的缩写，其语法如下：

```
flex: none | flex-grow flex-shrink flex-basis
```

其中第二个参数和第三个参数（flex-shrink和flex-basis）是可选参数。默认值为0 1 auto。

HTML代码如下：

```html
<span class="flex-container">
  <span class="flex-item item1" >1</span>
  <span class="flex-item item2" >2</span>
  <span class="flex-item item3" >3</span>
</span>
```

示例CSS代码如下：

```css
.flex-container {
  display: flex;
  flex-direction: row;
  width: 200px;
  height: 150px;
}
.flex-item {
  width: 50px;
  height: 50px;
}
.item3 {
  flex: 1;
}
```

效果如图2-31所示。

说明　本例子中，类名为item3的元素宽度为50。当该元素被设置为flex:1后，该元素就会把伸缩容器的剩余空间占满，其实本质上就等于flex-grow:1。

该属性有两个快捷值：auto（即1 1 auto）和none（即0 0 auto）。

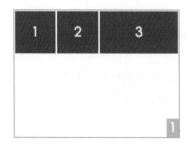

图2-31　flex为1的效果图

6. align-self

该属性用来设置单独的伸缩项目在交叉轴上的对齐方式,会覆写默认的对齐方式,其语法如下：

align-self: auto | flex-start | flex-end | center | baseline | stretch

HTML代码如下：

```
<span class="flex-container">
  <span class="flex-item item1" >1</span>
  <span class="flex-item item2" >2</span>
  <span class="flex-item item3" >3</span>
</span>
```

下面简要介绍这6个属性值。

❑ **auto**：伸缩项目按照自身设置的宽高显示,如果没有设置,则按stretch来计算其值。示例CSS代码如下：

```
.flex-container {
  display: flex;
  flex-direction: row;
  flex-wrap: wrap;
  width: 200px;
  height: 150px;
}
.flex-item {
  width: 50px;
  height: 50px;
}
.item3 {
  align-self: auto;
}
```

效果如图2-32所示。

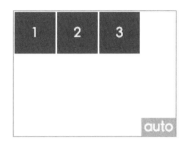

图2-32　align-self为auto的效果图

❑ **flex-start**：伸缩项目向交叉轴的开始位置靠齐。示例CSS代码如下：

```
.flex-container {
  display: flex;
  flex-direction: row;
  flex-wrap: wrap;
  width: 200px;
  height: 150px;
}
.flex-item {
  width: 50px;
  height: 50px;
}
.item3 {
  align-self: flex-start;
}
```

效果如图2-33所示。

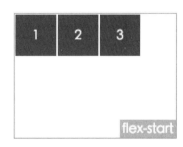

图2-33　align-self为flex-start的效果图

❑ **flex-end**：伸缩项目向交叉轴的结束位置靠齐。示例CSS代码如下：

```
.flex-container {
  display: flex;
  flex-direction: row;
  flex-wrap: wrap;
  width: 200px;
  height: 150px;
}
```

```
.flex-item {
  width: 50px;
  height: 50px;
}
.item1 {
  align-self: flex-end;
}
```

效果如图2-34所示。

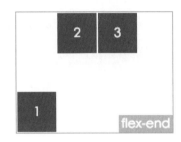

图2-34　align-self为flex-end的效果图

❑ **center**：伸缩项目向交叉轴的中心位置靠齐。示例CSS代码如下：

```
.flex-container {
  display: flex;
  flex-direction: row;
  flex-wrap: wrap;
  width: 200px;
  height: 150px;
}
.flex-item {
  width: 50px;
  height: 50px;
}
.item1 {
  align-self: center;
}
```

效果如图2-35所示。

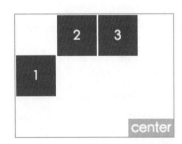

图2-35　align-self为center的效果图

❑ **baseline**：伸缩项目按基线对齐。示例CSS代码如下：

```
.flex-container {
  display: flex;
  flex-direction: row;
  flex-wrap: wrap;
  width: 200px;
  height: 150px;
}
.flex-item {
  width: 50px;
  height: 50px;
}
.item1 {
  align-self: baseline;
  font-size: 40px;
}
.item2 {
  align-self: baseline;
  font-size: 30px;
}
.item3 {
  align-self: baseline;
  font-size: 20px;
}
```

效果如图2-36所示。

图2-36　align-self为baseline的效果图

> **说明**　从示例中可以看到，每个伸缩项目都是沿交叉轴的方向，按照前一个伸缩项目的基线作为对齐起点的。

❑ **stretch**：伸缩项目在交叉轴方向占满伸缩容器。示例CSS代码如下：

```
.flex-container {
  display: flex;
  flex-direction: row;
  flex-wrap: wrap;
  width: 200px;
```

```
      height: 150px;
}
.flex-item {
  width: 50px;
}
.item1 {
  align-self: stretch;
}
.item2 {
  align-self: stretch;
  height: 50px;
}
.item3 {
  height: 50px;
}
```

效果如图2-37所示。

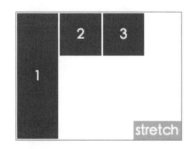

图2-37　align-self为stretch的效果图

> **说明**　本例中交叉轴为垂直轴，所以只有在不设置高度的情况下才能看到效果。在类名为item2的元素中，虽然设置了stretch，但是由于设置了height，所以看不到效果。

2.1.5　在 React Native 中使用 flexbox

React Native将Web中的flexbox布局引入进来，使得视图的布局更加简单。从官网上了解到，React Native目前主要支持flexbox的如下属性：

- alignItems
- alignSelf
- flex
- flexDirection
- flexWrap
- justifyContent

下面简要介绍这6个属性。

1. alignItems

该属性的用法同前面的align-items，区别在于它需要用驼峰拼写法，其语法如下：

```
alignItems: flex-start | flex-end | center | stretch
```

2. alignSelf

该属性的用法同上面的align-self，区别在于它需要用驼峰拼写法，其语法如下：

```
alignSelf: auto | flex-start | flex-end | center | stretch
```

3. flex

该属性的用法同上面的flex，其语法如下：

```
flex: number
```

4. flexDirection

该属性的用法同上面的flex-direction，区别在于React Native中默认的是column，其语法为：

```
flexDirection: row | column
```

5. flexWrap

该属性的用法同上面的flex-wrap，区别在于需要用驼峰拼写法，其语法如下：

```
flexWrap: wrap | nowrap
```

6. justifyContent

该属性的用法同上面的justify-Content，区别在于需要用驼峰拼写法，其语法如下：

```
justifyContent: flex-start | flex-end | center | space-between | space-around
```

2.1.6 实例

我们用flexbox布局来实现一个标准的盒子模型应用BoxApp。对于盒子模型，用过的同学应该都比较熟悉，它是CSS中排版布局的重要方法。它包括的属性有margin、border和padding等。

这里的运行环境为iPhone 6 Plus，我们先从一个简单的容器壳开始。首先列出HTML5中的主要代码实现。

HTML代码如下：

```html
<span class="margginBox">
  <span class="box height400 width400">
    <span class="label">margin</span>
    <span class="top height50 bgred">top</span>
    <span class="borderBox">
```

```html
      <span class="left bgred">left</span>
      <span class="right bgred">right</span>
    </span>
    <span class="bottom height50 bgred">bottom</span>
  </span>
</span>
```

CSS代码如下：

```css
.margginBox {
  position: absolute;
  top: 100px;
  padding-left: 7px;
  padding-right: 7px;
}
.box {
  display: flex;
  flex-direction: column;
  flex: 1;
  position: relative;
  color: #FDC72F;
  line-height: 1em;
}
.height400 {
  height: 400px;
}
.width400 {
  width: 400px;
}
.label {
  top: 0;
  left: 0;
  padding: 0 3px 3px 0;
  position: absolute;
  background-color: #FDC72F;
  color: white;
  line-height: 1em;
}
.top {
  width: 100%;
  justify-content: center;
  display: flex;
  align-items: center;
}
.bottom {
  width: 100%;
  display: flex;
  justify-content: center;
  align-items: center;
}
.right {
  width: 50px;
  display: flex;
  justify-content: space-around;
```

```
    align-items: center;
}
.left {
  width: 50px;
  display: flex;
  justify-content: space-around;
  align-items: center;
}
.height50 {
  height: 50px;
}
.bgred {
  background-color: #6AC5AC;
}
```

下面我们来分析其中主要的CSS代码。

- `.margginBox`定义了一个容器的位置，该容器距top为100px。
- `.box`定义为一个伸缩容器，且以交叉轴方向伸缩。
- `.height400`定义了高度为400px。
- `.height50`定义了高度为50px。
- `.width400`定义了宽度为400px。
- `.bgred`定义背景颜色。
- `.label`定义标签的显示位置，本例中是以绝对定位的方式显示在容器的左上角。
- `.top`、`.bottom`定义了宽度为100%，也定义自身为伸缩容器，并设置了内部元素沿主轴和交叉轴都居中。
- `.right`、`.left`定义了宽度为50px，也定义自身为伸缩容器，并设置了内部元素沿交叉轴居中，沿主轴等间隔分布。

下面看一下React Native中的主要代码实现：

```
var React = require('react-native');
var {
  AppRegistry,
  StyleSheet,
  Text,
  View,
} = React;
var BoxStyles = StyleSheet.create({
  "height50": {
    height: 50,
  },
  "height400": {
    height: 400,
  },
  "width400": {
    width: 400,
  },
  "bgred": {
```

```
      backgroundColor: "#6AC5AC",
    },
    "box": {
      flexDirection: "column",
      flex: 1,
      position: "relative",
    },
    "label": {
      top: 0,
      left: 0,
      paddingTop: 0,
      paddingRight: 3,
      paddingBottom: 3,
      paddingLeft: 0,
      position: "absolute",
      backgroundColor: "#FDC72F",
    },
    "top": {
      justifyContent: "center",
      alignItems: "center",
    },
    "bottom": {
      justifyContent: "center",
      alignItems: "center",
    },
    "right": {
      width: 50,
      justifyContent: "space-around",
      alignItems: "center",
    },
    "left": {
      width: 50,
      justifyContent: "space-around",
      alignItems: "center",
    },
    "yellow": {
      color: "#FDC72F",
      fontWeight:"900",
    },
    "white": {
      color: "white",
      fontWeight:"900",
    },
    "margginBox":{
      "position": "absolute",
      "top": 100,
      "paddingLeft":7,
      "paddingRight":7,
    },
    "borderBox":{
      flex: 1,
      justifyContent: "space-between",
      flexDirection: "row",
    }
```

```
  })
  var BoxContainer = React.createClass({
    render:function(){
      return (
        <View style={[BoxStyles.margginBox]}  ref="lab1">
          <View style={[BoxStyles.box,BoxStyles.height400,BoxStyles.width400]}>
            <View style={[BoxStyles.top,BoxStyles.height50,BoxStyles.bgred]}>
              <Text style={BoxStyles.yellow}>top</Text>
            </View>
            <View style={[BoxStyles.borderBox]}>
              <View style={[BoxStyles.left,BoxStyles.bgred]} >
                <Text style={BoxStyles.yellow}>left</Text>
              </View>
              <View style={[BoxStyles.right,BoxStyles.bgred]}>
                <Text style={BoxStyles.yellow}>right</Text>
              </View>
            </View>
            <View style={[BoxStyles.bottom,BoxStyles.height50,BoxStyles.bgred]}>
              <Text style={BoxStyles.yellow}>bottom</Text>
            </View>
            <View style={[BoxStyles.label]} >
              <Text style={BoxStyles.white}>margin</Text>
            </View>
          </View>
        </View>
      )
    }
  })
  AppRegistry.registerComponent('Box', () => BoxContainer);
```

可以看到，在React Native中，全部的代码都是由JavaScript来完成的。

图2-38是marginBox容器的效果。

图2-38　BoxApp之marginBox容器效果图

在上面的示例中，HTML5代码中涉及的属性如下：

- display
- flex-direction
- flex
- justify-content
- align-items

React Native中涉及的属性如下：

- justifyContent
- alignItems
- flex
- flexDirection

下面我们分别从样式、元素和书写格式这3个方面简要介绍HTML5与React Native的差异。

- **样式**。HTML5上的写法为：

```
.margginBox {
  position: absolute;
  top: 100px;
  padding-left: 7px;
  padding-right: 7px;
}
```

React Native上的写法为：

```
"margginBox":{
  "position": "absolute",
  "top": 100,
  "paddingLeft": 7,
  "paddingRight": 7,
}
```

- **元素**。HTML5代码如下：

```
<span class="margginBox">
  <span class="box height400 width400>
  <span class="label">margin</span>
  ……
```

而React Native代码如下：

```
……
    <View style={[BoxStyles.label]} >
      <Text style={BoxStyles.white}>margin</Text>
    </View>
  </View>
</View>
```

可以看出，HTML5上元素的显示层级是绝对定位层级较高，而React Native中的显示层级是后面比前面的高。若在React Native中将代码提前到第一行，它将会被后面的元素覆盖掉。

- **书写格式**。这方面的差异如下所示。

  HTML5以分号（；）结尾，React Native以逗号（,）结尾。

  HTML5中key、value都不加引号，React Native中属于JavaScript对象，如果value为字符串，需要用引号引起来，并且key值中间不能出现" - "，一定要转为驼峰命名方式。

  HTML5中，value值为数字时，需带单位，React Native中则不用带。

案例的全部代码在后面的案例代码库中可以下到，最终的效果如图2-39所示。

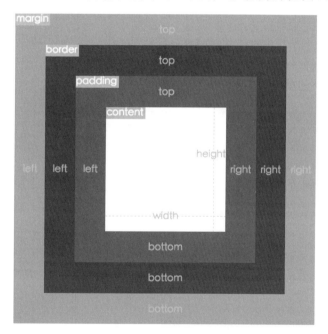

图2-39　BoxApp效果图

2.2　React 中的 JSX

首先，我们先认识下React。React 由ReactJS与React Native组成，其中ReactJS是Facebook开源的一个前端框架，React Native是ReactJS思想在native上的体现。

2.2.1　JSX 入门

我们先从hello world开始，下面是基于JSX写的一段代码：

```
var component = React.createClass({
  render: function() {
    return <div>hello world</div>
  }
});
var component = React.createClass({
  render: function() {
    return  React.createElement('div', null, 'hello world')
  }
});
```

代码给人最直观的感觉就是，对于习惯了Web开发的人来说，在JavaScript代码中写HTML代码这种做法刚开始可能比较难接受。但是当你习惯后，便会觉得它不是那么别扭，反而很舒服，它带来的最直接的好处便是能够很直观地构建出页面结构。

那么，什么是JSX呢？其实JSX并不是一门新的语言，仅仅是个语法糖，允许开发者在JavaScript中书写HTML语法。最后，每个HTML标签都转化为JavaScript代码来运行。这样对于使用JavaScript来构建组件以及组件之间关系的应用，在代码层面显得更加清晰，而不再是用JavaScript操作DOM来创建组件以及组件之间的嵌套关系。

> **说明** 语法糖（syntactic sugar），也译为糖衣语法，是由英国计算机科学家彼得·约翰·兰达（Peter J. Landin）发明的一个术语，指计算机语言中添加的某种语法，这种语法对语言的功能并没有影响，但是更方便程序员使用。通常来说，使用语法糖能够增加程序的可读性，从而减少程序代码出错的机会。

1. 环境

JSX必须借助ReactJS环境才能运行，在编写JSX代码之前，需要先加载ReactJS文件，如：react-0.13.3.js。当然光有ReactJS还不行，还需要加载JSX的解析器，如：JSXTransformer-0.13.3.js。所以最终的运行环境代码为：

```
<script src="./build/react-0.13.3.js"></script>
<script src="./build/JSXTransformer-0.13.3.js"></script>
```

2. 载入方式

JSX目前有两种方法载入。

❏ **内联方式的载入**。相关代码如下：

```
<script type="text/jsx">
  //这里书写JSX代码
  React.render(
    <h1>Hello, world!</h1>,//这便是内联的JSX代码
    document.getElementById('example')
  );
</script>
```

❑ **外联方式的载入**。将下面的代码单独放在一个helloworld.jsx的文件中：

```
helloworld.jsx:
React.render(
  <h1>Hello, world!</h1>,
  document.getElementById('example')
);
```

然后在页面上通过下面的方式引入helloworld.jsx文件：

```
<script type="text/jsx" src="helloworld.jsx"></script>
```

3. 标签

JSX标签其实就是HTML标签，如：

```
<h1>Hello, ReactJS!!</h1>
```

只是我们在JavaScript中书写这些标签时，不再像以前那样作为字符串用引号引起来，而是像在XML文件中书写一样，直接写即可。然而还有一类标签是HTML所没有的，但是我们也可以使用，那就是用ReactJS创建的组件类标签，示例代码如下：

```
var Hello = React.createClass({
  render: function() {
    return <div>Hello</div>;
  }
});
React.render(
  <Hello/>,
  document.getElementById('container')
)
```

在上面的代码中，我们创建了一个叫Hello的组件，此时我们就可以像使用HTML标签一样，通过<Hello/>的方式把它引入进来。不过有一点需要注意的是，ReactJS中约定自定义的组件标签首字母一定要大写，这样便于区分是组件标签还是HTML标签。

4. 转换

JSX的书写，只是为了让我们能更直观地看到组件的DOM结构，它并不能直接在浏览器端运行。最后只能通过解析器将其转化为JavaScript代码才能在浏览器端运行。

比如，我们写了一段代码：

```
<h1>Hello, ReactJS!! </h1>
```

那么解析器会将其转化为

```
React.createElement("h1", null, "Hello, ReactJS!!")
```

可以看到，其实我们每写一个标签，就相当于调用一次React.createElement方法并最后返回一个ReactElement对象给我们。

那么，React.createElement方法中的这些参数分别代表什么意思呢？下面我们看官方的一个参数解释：

```
React.createElement(
  string/ReactClass type,
  [object props],
  [children ...]
)
```

该方法的第一个参数可以是一个字符串，表示是一个HTML标准内的元素，或者是一个ReactClass类型的对象，表示我们之前封装好的自定义组件。第二个参数是一个对象，或者说是字典也可以，它保存了这个元素的所有固有属性（即传入后基本不会改变的值）。从第三个参数开始，之后的参数都被认作是元素的子元素。

5. 执行JavaScript表达式

做Web开发时，经常会用到模板。我们知道模板中是可以执行JavaScript代码的，那么JSX也不例外。接下来，我们看下JSX是怎么执行JavaScript代码的。

下面我们先定义一个变量：

```
var msg = "Hello ReactJS!!";
```

然后在JSX中调用该变量：

```
<h1>{msg}</h1>
```

解析完后的结果如下：

```
var msg = "Hello, ReactJS!!";
React.createElement("h1", null, msg)
```

可以看出，在JSX中运行JavaScript表达式，需要将表达式用 {} 括起来。

6. 注释

书写代码的时候，注释是必不可少的，一来便于了解代码的结构，二来便于日后的维护。在JSX中，当然也可以使用注释，那么它的注释方式又是如何的呢？

其实我们可以猜到，JSX最终都是被编译为JavaScript运行的，所以它的注释应该和JavaScript一样。没错，你的猜测是正确的。

在JSX中，也分为单行注释和多行注释。下面我们通过一个示例来更好地理解它。

这是书写的JSX代码：

```
var msg = "Hello, ReactJS!!";
//<h1>msg</h1>单行注释
/**
  <h1>msg</h1>多行注释
**/
```

这是编译后的JSX代码：

```
var msg = "Hello, ReactJS!!";
//<h1>msg</h1>单行注释
/**
   <h1>msg</h1>多行注释
**/
```

7. 属性

我们知道，在HTML中可以通过标签上的属性来改变当前元素的样式。当然，在JSX中也可以使用该功能。我们看下面的一个示例：

```
var msg = <h1 width="10px">Hello, ReactJS!!</h1>;
```

这里我们设置了h1标签的宽度为10px，该行代码转化后的结果为：

```
var msg = React.createElement("h1", {width: "10px"}, "Hello, ReactJS!!");
```

可以看到，我们设置的属性都被作为React.createElement方法的第二个参数输入进去了。这也对应了上面对React.createElement第二个参数的介绍。

8. 延展属性

随着JavaScript的飞速发展，ES6的功能也逐渐被浏览器厂商所支持。在JSX中，我们完全可以使用ES6的语法。因为我们有强大的解析器会对不支持ES6语法的浏览器做降级处理。我们看下面的示例：

```
var props = {};
props.foo ="1";
props.bar = "1";
<h1 {...props} foo="2" >Hello, ReactJS!!</h1>
```

上面的书写会被转化为：

```
var props = {};
props.foo ="1";
props.bar = "1";
React.createElement("h1", React.__spread({}, props, {foo: "2"}), "Hello, ReactJS!!")
```

这里我们最关注的便是"..."，这是什么用法呢？在ES6中，该功能是用来遍历对象的，那么上面所写的...props的意思便是遍历props这个对象，并且将它的所有属性都赋值给h1这个元素。这里要注意的是，元素后面定义的属性会覆盖掉前面的属性。上例中的foo="2"便把之前的foo ="1"覆盖了。

9. 自定义属性

之前我们讲到的"属性"是指标签特有的属性，自定义属性便是自己定义的一些属性，但是在真正的页面渲染后，它不一定显示在页面上。那么，什么样的自定义属性才能渲染在页面上呢？

HTML5给出了方案，凡是以data-开头的自定义属性，在页面渲染后均可以显示在页面上。为了更加清晰易懂，我们来看个例子。

我们在JSX中进行如下的书写：

```
var hello = <h1 height="100" data-test="test" test="test">Hello, ReactJS!!!</h1>;
React.render(
  hello,
  document.getElementById('example')
);
```

经过JSX解析器解析后的结果为：

```
var hello = React.createElement("h1", {height: "100",
  "data-test": "test", test: "test"}, "Hello, ReactJS!!!");
React.render(
  hello,
  document.getElementById('example')
);
```

现在看最后渲染在页面上的显示结果：

```
<h1 height="100" data-test="test" data-reactid=".0">Hello, ReactJS!!!</h1>
```

在该例子中，我们自定义了属性data-test和test，最后页面中只显示了data-test。可见，以data-开始的自定义属性会被渲染到页面上，其他自定义属性会被忽略掉。

10. 显示HTML

有时候，我们需要显示一段HTML字符串，而不是HTML节点，那么这在JSX中怎么实现呢？这需要我们的_html属性上场了。我们看下面的例子：

```
<div>{{_html:'<h1>Hello, ReactJS!!!</h1>'}}</div>
```

转化后的结果为：

```
React.createElement("div", null, {_html:'<h1>Hello, ReactJS!!!</h1>'})
```

此时JSX就以字符串的方式将HTML显示在了页面上。

11. 样式使用

在做Web开发中，元素的样式是必不可少的角色，那么在JSX中能像在HTML中那样使用样式吗？答案是否定的。那么，在JSX中究竟该如何使用样式呢？这与我们用JavaScript直接操作样式很类似。下面是一段在JSX中使用样式的代码。

JSX中的输入：

```
<h1 style={{color: '#ff0000', fontSize: '14px'}}>Hello, ReactJS!!!</h1>
```

解析器转化后的结果：

```
React.createElement("h1", {style: {color: '#ff0000', fontSize: '14px'}}, "Hello, ReactJS!!!")
```

下面我们分析一下上面的代码。首先，JSX中的样式是通过style属性定义的。和传统Web定义不同的是，它不再是一个字符串而是一个JavaScript对象。在上面的代码中，第一个大括号是JSX语法，第二个大括号是JavaScript对象，我们把需要定义的样式都以对象的方式写在这个大括号里。需要注意的是，我们要将属性名转为驼峰命名格式，若不转的话，需要加引号括起来。比如，'font-size':'13px'或者fontSize:'13px'，这其实就是JavaScript中JSON的书写方式。当然，我们也可以通过className=xxx的方式引入样式，但是切记是className而不是class。

12. 事件绑定

通过上面的学习，我们了解到JSX中结构的定义和样式的使用。接下来，我们主要说说JSX中的交互用法。我们知道没有交互，就相当于没有灵魂。那么，JSX中的交互到底是什么样子呢？

这是一段JSX中事件绑定的代码：

```
function testClick(){
  alert('testClick');
}
var app= <button  onClick={testClick.bind(this)} style={{
    bacogroundColor: '#ff0000',
    fontSize: '28px',
    width:'200px',
    height:'100px'
  }}>Hello, ReactJS!!!</button>
React.render(
  app,
  document.getElementById('example')
);
```

这段代码实现的功能是，点击一个按钮的时候，弹出一个写着testClick的提示框。这个功能看起来和在HTML中使用的绑定事件几乎一样。没错，JSX支持所有的HTML元素的事件。不过有几点还是需要拿出来提醒一下大家。首先，事件名称一定要采用驼峰命名方式，上面的onClick要是换成onclick就不起效果了。其次，有个技巧给大家分享下，如果要给绑定的事件传递参数，若要给testClick传递一个hello的字符串，可以这样写onClick={testClick.bind(this,'hello')}。可以看到，这里需要调用bind方法，这个方法的第一个参数主要用来设置作用域，从第二个开始便是我们想要传递的参数了。

通过上面的学习，我们已经对JSX有了一个很全面的认识，可能有的同学已经开始按捺不住，想要开始写点什么啦，下面我们就来讲讲JSX的实际应用场景。

2.2.2　JSX 实战之 ReactJS

说到JSX在ReactJS中的用法，我们需要先了解ReactJS中的一些概念。了解这些概念，对于我们掌握JSX的使用方法很有帮助。下面我们主要从ReactJS的组件入手，带领大家了解ReactJS，了解JSX在其中的使用情况。

1. ReactJS 简介

ReactJS的核心思想便是组件化，即按功能封装成一个一个的组件，各个组件维护自己的状态和UI，当状态发生变化时，会自动重新渲染整个组件，多个组件一起协作共同构成了ReactJS应用。

2. 组件介绍

我们先从一个简单的组件开始：

```
var HelloMessage = React.createClass({
  render: function() {
    return <div>Hello {this.props.name}</div>;
  }
});
React.render(
  <HelloMessage name="John" />,
  document.getElementById('example')
)
```

这是一个简单的HelloMessage组件，其功能是接收传入的用户名称，然后在页面中显示Hello John。从代码中可以看到，这里主要用到了React对象，以及React.createClass、render和React.render方法。下面我们来逐一解释一下它们。

- React是全局对象，ReactJS的所有顶层API都挂在这个对象下。
- React.createClass是ReactJS用来创建组件类的方法。
- render是一个组件级的API，是React.createClass内部最常用的API，该方法主要用来返回组件结构。
- React.render是最基本的方法，用于将模板转为HTML语言，并将转化后的DOM结构插入到指定的DOM节点上。该方法的参数描述如下：

```
ReactComponent render(
  ReactElement element,//ReactJS组件
  DOMElement container,//页面中的DOM容器
  [function callback]//插入后的回调
)
```

如果开头的代码用文字描述的话，那就是先通过React.createClass来创建一个组件类，然后调用render方法输出组件的DOM结构，最后调用React.render将组件插入在id为example的节点上。

在ReactJS中，从大的分类来看，主要分为顶层API（即挂在React对象下的API，如React.createClass、React.render等）和组件API（只有组件内部才可以调用的API，如render）。这里我们只简单介绍了一些基本常用的API，要了解详细的API，可以访问http://facebook.github.io/react/docs/getting-started.html。

当用React.createClass创建组件类时，除了要包含一个render方法外，也可以包含组件生命

周期中的其他方法。

下面我们来学习一下组件的生命周期。

3. 组件的生命周期

这里我们通过构建一个简单的列表组件，来深入了解组件各环节的运作流程。下面是该列表组件的详细代码：

```javascript
var List = React.createClass({
  //1.创建阶段
  getDefaultProps: function() {
    //在创建类的时候被调用
    console.log("getDefaultProps");
    return {};
  },
  //2.实例化阶段
  getInitialState: function() {
    //获取this.state的默认值
    console.log("geyInitialState");
    return {};
  },
  componentWillMount: function() {
    //在render之前调用此方法
    //业务逻辑的处理都应该放在这里，如对state的操作等
    console.log("componentWillMount");
  },
  render: function() {
    //渲染并返回一个虚拟DOM
    console.log("render");
    return (
      <div> hello <strong> {this.props.name} </strong></div>
    );
  },
  componentDidMount: function() {
    //该方法发生在render方法之后。在该方法中，ReactJS会使用render方法返回的虚拟DOM对象来创建真实的
    //DOM结构
    console.log("componentDidMount");
  },
  //3. 更新阶段
  componentWillRecieveProps: function() {
    //该方法发生在this.props被修改或父组件调用setProps()方法之后
    console.log("componentWillRecieveProps");
  },
  shouldComponentUpdate:function(){
    //是否需要更新
    console.log("shouldComponentUpdate");
    return true;
  },
  componentWillUpdate: function() {
    //将要更新
    console.log("componentWillUpdate");
```

```
    },
    componentDidUpdate: function() {
      //更新完毕
      console.log("componentDidUpdate");
    },
    //4. 销毁阶段
    componentWillUnmount:function(){
      //销毁时被调用
      console.log("componentWillUnmount");
    }
});
```

在整个ReactJS的生命周期中，主要会经历这4个阶段：创建阶段、实例化阶段、更新阶段和销毁阶段。

❑ **创建阶段**。该阶段主要发生在创建组件类的时候，即在调用React.createClass的时候。这个阶段只会触发一个getDefaultProps方法，该方法会返回一个对象，并缓存下来。然后与父组件指定的props对象合并，最后赋值给this.props作为该组件的默认属性。

这里我们说明一下props。它是一个对象，是组件用来接收外面传来的参数的，组件内部是不允许修改自己的props属性的，只能通过父组件来修改。上面的getDefaultProps方法便是处理props的默认值的。

❑ **实例化阶段**。该阶段主要发生在实例化组件类的时候，也就是该组件类被调用的时候。下面我们来实例化一个List类：

```
React.render(
  <List name = 'ReactJS' > </List>,
  document.getElementById("container")
);
```

即List组件类被调用的时候。这个阶段会触发一系列的流程，具体的执行顺序如下所示。

(1) getInitialState。初始化组件的state的值，其返回值会赋值给组件的this.state属性。

(2) componentWillMount。根据业务逻辑来对state进行相应的操作。

(3) render。根据state的值，生成页面需要的虚拟DOM结构，并返回该结构。

(4) componentDidMount。对根据虚拟DOM结构而生成的真实DOM进行相应的处理，组件内部可以通过this.getDOMNode()来获取当前组件的节点。然后就可以像在Web开发中那样操作里面的DOM元素了。

上面我们提到了几个陌生的术语——state和虚拟DOM，这里简要解释一下它们。

state：它是组件的属性，主要用来存储组件自身需要的数据。它是可以改变的，它的每次改变都会引发组件的更新，这也是ReactJS中的关键点之一。每次数据的更新都是通过修改state属性的值，然后ReactJS内部会监听state属性的变化，一旦发生变化，就会主动触发组件的render方法来更新DOM结构。

虚拟DOM：它是ReactJS中提出的一个概念，是将真实的DOM结构映射成一个JSON数据结构。

- **更新阶段**。这主要发生在用户操作之后或者父组件有更新的时候，此时会根据用户的操作行为进行相应的页面结构的调整。该阶段也会触发一系列的流程，具体的执行顺序如下所示。

  (1) componentWillReceiveProps(object nextProps)。当组件接收到新的props时，会触发该函数。在该函数中，通常可以调用this.setState方法来完成对state的修改。

  (2) shouldComponentUpdate(nextProps, nextState)。该方法用来拦截新的props或state，然后根据事先设定好的判断逻辑，做出最后要不要更新组件的决定。

  (3) componentWillUpdate(object nextProps, object nextState)。当步骤(2)的shouldComponentUpdate方法中的拦截返回true的时候，就可以在该方法中做一些更新之前的操作。

  (4) render。根据一系列的diff算法，生成需要更新的虚拟DOM数据。实践表明，在render中，最好只做数据和模板的组合，不应进行state等逻辑的修改，这样组件结构更清晰。

  > **说明** 当返回null或者false时，表明不需要渲染任何东西。

  (5) componentDidUpdate。该方法在组件的更新已经同步到DOM中后触发，我们常在该方法中做一些DOM操作。

- **销毁阶段**。既然说的是生命，那么有生就有死，组件也不例外。这里我们主要讲组件死的时候会触发的方法，目前来看只会触发一个叫componentWillUnmount的方法。当组件需要从DOM中移除的时候，我们通常会做一些取消事件绑定、移除虚拟DOM中对应的组件数据结构、销毁一些无效的定时器等工作，这些事情都可以在这个方法中处理。

4. 组件之间的通信

之前已经提到过，ReactJS的思想是组件化，现在我们便可以尝试着做一个简单的应用。

先创建一个父类组件Parent，它内部调用的是一个叫Child的子组件，并将接收到的外部参数name传递给子组件Child。先创建父类组件Parent：

```
var Parent = React.createClass({
  click:function(){
    this.refs.child.getDOMNode().style.color="red";
  },
  render: function() {
    return (
      <div onClick={this.click}>
        Parent is:
        <Child name = {this.props.name} ref="child"> </Child>
      </div>
```

```
    );
  }
});
```

然后创建子类组件Child：

```
var Child = React.createClass({
  render: function() {
    return <span> {this.props.name} </span>;
  }
});
```

最后，通过React.render方法将组件渲染在页面上：

```
React.render( <Parent name='React' /> ,document.getElementById('container'));
```

整个应用的功能是父组件接收传入的用户名称，并将用户名称传给子组件，最后再由子组件渲染出来显示在页面上。

可以看到，这里组件之间的组合关系是一层套着一层，就像HTML中的DOM结构一样，所以用ReactJS开发的应用，在组织结构方面很清晰明了。

现在回到我们的标题，这些组件之间是怎么通信的呢，下面我们从两个方面来介绍一下。

- **子组件调用父组件的方法**。从上面的代码中可以看到，子组件要拿到父组件中的属性，需要通过this.props方法。没错，就是这个方法打通了子组件与父组件的通信桥梁。所以，如果子组件想要调用父组件的方法，只需父组件把要被调用的方法以属性的方式放在子组件上，子组件内部便可以通过"this.props.被调用的方法"这样的方式来获得父组件传过来的方法。代码中的this.props.name便是Child组件从Parent组件拿到name属性的。然后，每次父组件修改了传入的name属性，子组件便会得到通知，然后会自动获取新的name属性。
- **父组件调用子组件的方法**。子组件调用父组件是通过props实现的，那么如何实现父组件调用子组件呢？在ReactJS中，有个很巧妙的属性，那就是ref。这个属性就像给组件起个名字一样，子组件被设置为ref之后，父组件便可以通过this.ref.xxx来获取到子组件了。其中xxx为子组件ref的值。细心的同学可能很早就注意到上述代码中的ref="child"了，这就是Parent组件为了获取Child组件而专门设置的。上面代码的意思是，当父组件被点击的时候，会先获取子组件元素，然后将子组件元素中的文字颜色改为红色。

5. 虚拟DOM

前面我们主要从大的方面掌握了组件的一些知识，接下来从细微的组件更新来说说ReactJS的精妙之处。说到更新，ReactJS最大的亮点就是Virtual DOM技术。那么，Virtual DOM到底是什么呢？之前我们了解到，当组件的state属性发生变化时，会自动执行组件的render方法来实现组件更新。但是如果真的这样大面积操作DOM的话，性能肯定会是一个很大的问题。然而聪明的ReactJS实现了Virtual DOM技术，将组件的DOM结构映射到这个Virtual DOM对象上，并且ReactJS还实现了一套Diff算法。当需要更新组件的时候，会通过Diff算法找到要变更的内容。最后，再把这个修改更新到实际的DOM节点上。所以，组件更新实际上不是真的渲染整个DOM树，而是只更新需要修改的DOM节点，这样在性能上会比原生DOM快很多。

这样描述可能有点抽象，下面看个具体的例子。

假设我们要创建一个组件，其结构如下：

```
<ul>
  <li>
    ctrip
  </li>
  <li>
    <ul>
      <li>
        elong
      </li>
    </ul>
  </li>
</ul>
```

通过前面的学习，我们的创建方式可能会有下面几种。

原生组件的创建方式

用JSX实现的代码如下：

```
var root = <ul>
  <li>
    ctrip
  </li>
  <li>
    <ul>
      <li>
        elong
      </li>
    </ul>
  </li>
</ul>

//主要用来输出虚拟DOM结构
console.log(root);
```

用JavaScript实现的代码如下：

```
var ctrip = React.createElement('li', null, 'ctrip');
var elong = React.createElement('ul', null, React.createElement('li', null, 'elong'));
var root = React.createElement('ul', null, ctrip,elong);
//主要用来输出虚拟DOM结构
console.log(root);
```

在控制台捕获的日志如图2-40所示，它展示了ReactJS对于原生组件生成的Virtual DOM的详细层次结构。

```
▼Object {$$typeof: Symbol(react.element), type: "ul", key: null, ref: null, props: Object…}
    $$typeof: Symbol(react.element)
    _owner: null
     self: null
    _source: null
  ▶ _store: Object
    key: null
  ▼ props: Object
    ▼ children: Array[2]
      ▼ 0: Object
          $$typeof: Symbol(react.element)
          _owner: null
           self: null
          _source: null
        ▶ _store: Object
          key: null
        ▼ props: Object
            children: "ctrip"
          ▶ __proto__: Object
          ref: null
          type: "li"
        ▶ __proto__: Object
      ▼ 1: Object
          $$typeof: Symbol(react.element)
          _owner: null
           self: null
          _source: null
        ▶ _store: Object
          key: null
        ▼ props: Object
          ▼ children: Object
              $$typeof: Symbol(react.element)
              _owner: null
               self: null
              _source: null
            ▶ _store: Object
              key: null
            ▼ props: Object
              ▼ children: Object
                  $$typeof: Symbol(react.element)
                  _owner: null
                   self: null
                  _source: null
                ▶ _store: Object
                  key: null
                ▶ props: Object
                  ref: null
                  type: "li"
                ▶ __proto__: Object
              ▶ __proto__: Object
              ref: null
              type: "ul"
            ▶ __proto__: Object
          ▶ __proto__: Object
          ref: null
          type: "li"
        ▶ __proto__: Object
        length: 2
      ▶ __proto__: Array[0]
    ▶ __proto__: Object
    ref: null
    type: "ul"
  ▶ __proto__: Object
```

图2-40　浏览器原生组件对应的Virtual DOM图

自定义组件的创建方式

相关代码如下：

```
var Ctrip = React.createClass({
  render: function() {
    return (
      <ul>
        <li>ctrip</li>
        <li>
```

```
          {this.props.children}
        </li>
      </ul>
    );
  }
});
var Elong = React.createClass({
  render: function() {
    return (
      <ul>
        <li>elong</li>
      </ul>
    );
  }
});
var root = <Ctrip><Elong></Elong></Ctrip>
//主要用来输出虚拟DOM结构
console.log(root);
```

在控制台中捕获的日志如图2-41所示。

```
▼Object {$$typeof: Symbol(react.element), key: null, ref: null, props: Object, owner: null…}
    $$typeof: Symbol(react.element)
    owner: null
    self: null
    _source: null
  ▶ store: Object
    key: null
  ▼props: Object
    ▼children: Object
        $$typeof: Symbol(react.element)
        owner: null
        self: null
        source: null
      ▶ store: Object
        key: null
      ▼props: Object
        ▶ __proto__ : Object
        ref: null
      ▼type: function (props, context, updater)
          arguments: (...)
          caller: (...)
          displayName: "Elong"
          length: 3
          name: ""
        ▶prototype: ReactClassComponent
        ▶ __proto__ : function ()
        ▶<function scope>
        ▶ __proto__ : Object
      ref: null
    ▼type: function (props, context, updater)
        arguments: (...)
        caller: (...)
        displayName: "Ctrip"
        length: 3
        name: ""
      ▶prototype: ReactClassComponent
      ▶ __proto__ : function ()
      ▶<function scope>
      ▶ __proto__ : Object
```

图2-41　自定义组件对应的Virtual DOM图

图2-41展示ReactJS对于自定义组件生成的虚拟DOM的详细层次结构。仔细观察可以发现，它与图2-40的数据结构略有不同，比如多了displayName属性等。

通过上面的日志可以看到，不论用哪种方式创建，在控制台中输出的日志都是一个JavaScript对象，这就是我们所说的Virtual DOM对象。

接下来，我们看看Diff算法在ReactJS中的体现。

首先，使用浏览器中MutationObserver的功能，对页面上元素的改变进行监听：

```html
<script type="text/javascript">
  var MutationObserver = window.MutationObserver
  || window.WebKitMutationObserver
  || window.MozMutationObserver;
  var observeMutationSupport = !!MutationObserver;
  var changedNodes = [];
  if(observeMutationSupport){
    var observer = new MutationObserver(function(mutations) {
      changedNodes = mutations;
      mutations.forEach(function(mutation) {
        console.log(mutation);
      })
    });
    var  options = {
      'childList': true,//子节点的变动
      'attributes':true ,//属性的变动
      'characterData': true,//节点内容或节点文本的变动
      'subtree':true ,//所有后代节点的变动
      'attributeOldValue': true,//表示观察attributes变动时，是否需要记录变动前的属性值
      'characterDataOldValue':true //表示观察characterData变动时，是否需要记录变动前的值
    } ;
    observer.observe(document.body, options);
  }
</script>
```

接着把ReactJS组件的生命周期做一次封装，便于组件来调用：

```html
<script type="text/babel">
  function LifeCycle(name){
    var obj = {
      name:name
    }
    for(var n in Cycle){
      obj[n] = Cycle[n];
    }
    return obj;
  }
  var Cycle = {
    getInitialState: function(){
      console.log(this.name,'getInitialState');
      return {};
    },
    getDefaultProps: function(){
      console.log(this.name,'getDefaultProps');
      return {};
    },
    componentWillMount: function(){
      console.log(this.name,'componentWillMount');
    },
```

```
      componentDidMount: function(){
        console.log(this.name,'componentDidMount');
      },
      componentWillReceiveProps: function(){
        console.log(this.name,'componentWillReceiveProps');
      },
      shouldComponentUpdate: function(){
        console.log(this.name,'shouldComponentUpdate');
        return true;
      },
      componentWillUpdate: function(){
        console.log(this.name,'componentWillUpdate');
      },
      componentDidUpdate: function(){
        console.log(this.name,'componentDidUpdate');
      },
      componentWillUnmount: function(){
        console.log(this.name,'componentWillUnmount');
      }
    };
</script>
```

然后定义需要用到的组件：

```
<script type="text/babel">
  //定义Ctrip组件
  var Ctrip = React.createClass({
    mixins: [LifeCycle('Ctrip')],
    render: function() {
      console.log('Ctrip','render');
      return (
        <ul>
          <li>ctrip</li>
          <li>
            <ul>
              {this.props.children}
            </ul>
          </li>
        </ul>
      );
    }
  });
  //定义Elong组件
  var Elong = React.createClass({
    mixins: [LifeCycle('Elong')],
    render: function() {
      console.log('Elong','render');
      return (
        <li>elong</li>
      );
    }
  });
  //定义Qunar组件
  var Qunar = React.createClass({
```

```
    mixins: [LifeCycle('Qunar')],
    render: function() {
      console.log('Qunar','render');
      return (
        <li>qunar</li>
      );
    }
  });
  //定义Ly组件
  var Ly = React.createClass({
    mixins: [LifeCycle('Ly')],
    render: function() {
      console.log('Ly','render');
      return (
        <li>ly</li>
      );
    }
  });
</script>
```

最后，定义我们的主逻辑：

```
//实现主逻辑
console.log('-----first------------');
ReactDOM.render(
  <Ctrip><Elong></Elong><Qunar></Qunar></Ctrip>,
  document.getElementById('container')
);
setTimeout(function(){
  console.log('-----second------------');
  ReactDOM.render(
    <Ctrip><Elong></Elong><Ly></Ly><Qunar></Qunar></Ctrip>,
    document.getElementById('container')
  );
},1000)
```

为了能够更好地验证ReactJS的Diff算法，我们借助了ReactJS的组件生命周期和浏览器的MutationObserver功能。通过组件的生命周期，我们能够很直观地看到Diff算法的体现；通过MutationObserver功能，我们又能很好地验证组件生命周期的正确性。上面代码的逻辑实现的效果其实就是在Elong和Qunar这两个组件之间插入一个Ly组件，从结构变化上来看应该是如下的样子。

一开始的结构如下：

```
<ul>
  <li>
    ctrip
  </li>
  <li>
    <ul>
      <li>
```

```
      elong
    </li>
    <li>
      qunar
    </li>
  </ul>
 </li>
</ul>
```

1秒之后的结构如下:

```
<ul>
  <li>
    ctrip
  </li>
  <li>
    <ul>
      <li>
        elong
      </li>
      <li>
        ly
      </li>
      <li>
        qunar
      </li>
    </ul>
  </li>
</ul>
```

常规的做法就是将Elong和Qunar组件先删除，然后依次创建和插入Elong、Ly和Qunar组件。

现在我们来看看ReactJS是怎么做的呢，我们通过在浏览器的控制台中进行捕获，看到的日志如图2-42所示。

```
-----second-------------
Ctrip componentWillReceiveProps
Ctrip shouldComponentUpdate
Ctrip componentWillUpdate
Ctrip render
Elong componentWillReceiveProps
Elong shouldComponentUpdate
Elong componentWillUpdate
Elong render
Qunar componentWillUnmount
Ly getInitialState
Ly componentWillMount
Ly render
Qunar getInitialState
Qunar componentWillMount
Qunar render
Elong componentDidUpdate
Ly componentDidMount
Qunar componentDidMount
Ctrip componentDidUpdate
▼MutationRecord {}
  ▶addedNodes: NodeList[0]
   attributeName: null
   attributeNamespace: null
   nextSibling: null
   oldValue: null
  ▶previousSibling: li
  ▶removedNodes: NodeList[1]
  ▶target: ul
   type: "childList"
  ▶ proto : MutationRecord
▼MutationRecord {}
  ▶addedNodes: NodeList[1]
   attributeName: null
   attributeNamespace: null
   nextSibling: null
   oldValue: null
  ▶previousSibling: li
  ▶removedNodes: NodeList[0]
  ▶target: ul
   type: "childList"
  ▶ proto : MutationRecord
▼MutationRecord {}
  ▶addedNodes: NodeList[1]
   attributeName: null
   attributeNamespace: null
   nextSibling: null
   oldValue: null
  ▶previousSibling: li
  ▶removedNodes: NodeList[0]
  ▶target: ul
   type: "childList"
  ▶ proto : MutationRecord
▶ chargedNodes[0]["removedNodes"]
◎ [ <li data-reactid=".0.1.0.1">qunar</li>]
▶ chargedNodes[1]["addedNodes"]
◎ [ <li data-reactid=".0.1.0.1">ly</li>]
▶ chargedNodes[2]["addedNodes"]
◎ [ <li data-reactid=".0.1.0.2">qunar</li>]
▶
```

图2-42 Diff算法在组件生命周期中的体现图

从日志中可以看出，ReactJS的Diff算法的结果是，Elong组件不变，先将Qunar组件进行删除，然后再创建并插入Ly组件，最后再创建并插入Qunar组件。可见，这比我们常规的做法省去了对Elong组件的删除操作。这样其实并没有将Diff算法的作用发挥到极致。下面我们再将主逻辑代码稍作调整：

```
console.log('-----first------------');
ReactDOM.render(
```

```
    <Ctrip><Elong key="Elong"></Elong><Qunar key="Qunar"></Qunar></Ctrip>,
    document.getElementById('container')
);
setTimeout(function(){
    console.log('-----second------------');
    ReactDOM.render(
        <Ctrip><Elong key="Elong"></Elong><Ly key="Ly"></Ly><Qunar key="Qunar"></Qunar></Ctrip>,
        document.getElementById('container')
    );
},1000)
```

从代码中可以看到，主要的修改就是给每个组件加了一个key属性，此时再来运行一下代码，在控制台中看到的日志如图2-43所示。

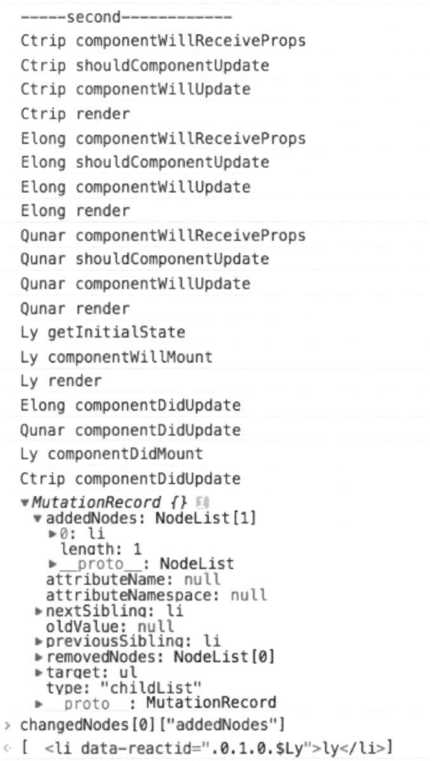

图2-43　Diff算法在添加key属性的组件生命周期中的体现图

可以看出，这次的Diff算法与之前有很大的不同。这次算法的结果是，Elong组件不变，Qunar组件不变，只是在Qunar组件之前创建并插入了Ly组件。

上面便是我们对Virtual DOM和Diff算法的一些简单使用和分析，如果大家还想更深入地了解Virtual DOM和Diff算法，可以去看http://calendar.perfplanet.com/2013/diff/这篇文章，里面有详细的讲解。

6. 实战

这里我们还是以上一节中的BoxApp为例进行介绍，首先看一下它的JSX实现。这里的运行环境是iPhone 6 Plus。

下面还是以创建一个容器壳开始，首先介绍一下HTML5中的主要代码实现。

HTML代码如下：

```
......
<div id="container"></div>
<script src="./build/react-with-addons-0.13.3.js"></script>
<script src="./build/JSXTransformer-0.13.3.js"></script>
......
```

CSS代码如下：

```
.height50 {
  height: 50px;
}
.height400 {
  height: 400px;
}
.width400 {
  width: 400px;
}
.bgred {
  background-color: #6AC5AC;
}
.box {
  display: flex;
  flex-direction: column;
  flex: 1;
  position: relative;
  color:#FDC72F;
  line-height: 1em;
}
.label {
  top: 0;
  left: 0;
  padding: 0 3px 3px 0;
  position: absolute;
  background-color: #FDC72F;
  color: white;
  line-height: 1em;
}
.top {
  width: 100%;
  justify-content: center;
  display: flex;
  align-items: center;
}
.bottom {
```

```css
  width: 100%;
  display: flex;
  justify-content: center;
  align-items: center;
}
.right{
  width: 50px;
  display: flex;
  justify-content: space-around;
  align-items: center;
}
.left {
  width: 50px;
  display: flex;
  justify-content: space-around;
  align-items: center;
}
.margginBox {
  position:absolute;
  top: 100px;
  padding-left:7px;
  padding-right:7px;
}
.borderBox {
  flex: 1;
  display: flex;
  justify-content: space-between;
}
```

JSX代码如下:

```jsx
var Box = React.createClass({
  render:function(){
    var parentClass = "box "+this.props.width+" "+ this.props.height;
    var topClass = "top height50 "+this.props.classBg;
    var leftClass = "left "+this.props.classBg;
    var rightClass = "right "+this.props.classBg;
    var bottomClass = "bottom height50 "+this.props.classBg;
    return (
      <span className= {parentClass}>
        <span className="label">{this.props.boxName}</span>
        <span className={topClass}>top</span>
        <span className={this.props.childName}>
          <span className={leftClass}>left</span>
            {this.props.children}
          <span className={rightClass}>right</span>
        </span>
        <span className={bottomClass}>bottom</span>
      </span>
    )
  }
})
var MargginBox = React.createClass({
  render:function(){
```

```
    return (
      <span className="margginBox">
        <Box  childName="borderBox"  height="height400" width="width400"
          boxName="margin" classBg="bgred">{this.props.children}</Box>
      </span>
    )
  }
})
var BoxContainer = React.createClass({
  render:function(){
    return (
      <MargginBox></MargginBox>
    )
  }
})
React.render(<BoxContainer/> ,document.getElementById("container"));
```

运行效果如图2-44所示。

图2-44　BoxApp之MarginBox容器图

下面我们分析一下JSX中的主要代码，这次从下往上看，这样逻辑会更加清晰。

- 最下面是调用React.render方法将BoxContainer组件渲染到ID为container的容器上。
- 在BoxContainer组件内部，只调用了MargginBox组件。
- MargginBox组件主要是调用Box组件，将height、width、boxName等值，以属性的方式传递给Box组件。要注意的一句代码是{this.props.children}，它会将MargginBox内部的子组件全部渲染出来。
- 在Box组件中，通过this.props来接收MargginBox组件传来的参数，通过className来给元素设置样式。这里也要强调一下{this.props.children}，这样Box组件就会把上面提到的MargginBox内部的子组件全部渲染到这里来。

全部代码可以在后面的案例代码库中下载，完整的效果图如图2-45所示。

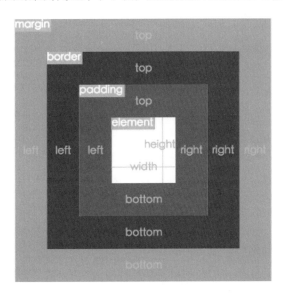

图2-45　BoxApp界面图

2.2.3　JSX 实战之 React Native

前面提到过，React Native是ReactJS思想在Native上的实现，也是通过组件化来构建应用的，而且组件的生命周期和ReactJS中的生命周期完全一样。不同的是，React Native中没有DOM的概念，只有组件的概念，所以我们在ReactJS中使用的HTML标签以及对DOM的操作在这里是用不了的。但是JSX的语法、事件绑定、自定义属性等这些特性，在React Native和ReactJS中是一样的。下面我们看看如何在React Native中实现上面的MarginBox容器，相关的JavaScript代码如下：

```
var React = require('react-native');
var {
  AppRegistry,
  StyleSheet,
  Text,
  View,
} = React;
var BoxStyles = StyleSheet.create({
  "height50": {
    height: 50,
  },
  "height400": {
    height: 400,
```

```
    },
    "width400": {
      width: 400,
    },
    "bgred": {
      backgroundColor: "#6AC5AC",
    },
    "box": {
      flexDirection: "column",
      flex: 1,
      position: "relative",
    },
    "label": {
      top: 0,
      left: 0,
      paddingTop: 0,
      paddingRight: 3,
      paddingBottom: 3,
      paddingLeft: 0,
      position: "absolute",
      backgroundColor: "#FDC72F",
    },
    "top": {
      justifyContent: "center",
      alignItems: "center",
    },
    "bottom": {
      justifyContent: "center",
      alignItems: "center",
    },
    "right": {
      width: 50,
      justifyContent: "space-around",
      alignItems: "center",
    },
    "left": {
      width: 50,
      justifyContent: "space-around",
      alignItems: "center",
    },
    "margginBox":{
      "position": "absolute",
      "top": 100,
      "paddingLeft":7,
      "paddingRight":7,
    }
})
var Box = React.createClass({
  render:function(){
    return (
      <View style={[BoxStyles.box,BoxStyles[this.props.width],
        BoxStyles[this.props.height]]}>
        <View style={[BoxStyles.top,BoxStyles.height50,BoxStyles[this.props.classBg]]}>
          <Text>top</Text></View>
```

```
        <View style={[BoxStyles[this.props.childName]]}>
          <View style={[BoxStyles.left,BoxStyles[this.props.classBg]]}>
            <Text>left</Text></View>
          {this.props.children}
          <View style={[BoxStyles.right,BoxStyles[this.props.classBg]]}>
            <Text>right</Text></View>
        </View>
        <View style={[BoxStyles.bottom,BoxStyles.height50,
          BoxStyles[this.props.classBg]]}><Text>bottom</Text></View>
        <View style={[BoxStyles.label]}><Text>{this.props.boxName}</Text></View>
      </View>
    )
  }
})
var MargginBox = React.createClass({
  render:function(){
    return (
      <View style={[BoxStyles.margginBox]}>
        <Box  childName="borderBox"  height="height400" width="width400"
          boxName="margin" classBg="bgred">{this.props.children}</Box>
      </View>
    )
  }
})
var BoxContainer = React.createClass({
  render:function(){
    return (
      <MargginBox>
      </MargginBox>
    )
  }
})
AppRegistry.registerComponent('Box', () => BoxContainer);
```

下面我们来分析一下上面的代码。

- 通过require导入react-native类库，并赋值给React变量。
- 从React变量导出我们需要的变量，如AppRegistry（用于注册组件）、StyleSheet（用于创建样式）、Text（文本组件，显示文字时使用）、View（视图组件，相当于HTML5中的div）等。
- 通过StyleSheet.create的方式创建BoxStyles样式实例。创建的时候，需要我们把事先定义好的JavaScript对象传入，在这个JavaScript对象中实现我们对样式的定义。
- 通过React.createClass的方式定义组件类。这里我们定义了Box、MargginBox和BoxContainer这3个组件，其中Box组件是核心，界面上显示的top、left、bottom、right等值都是在这个组件里面实现的。
- 将容器组件BoxContainer通过AppRegistry.registerComponent的方式注册到容器上。

上面简单介绍了MarginBox容器在React Native中的实现。关于整个案例的全部代码，可以在后面的案例代码库中下载到。

2.3 React Native 开发向导

通过前面的学习，我们已经掌握了开发React Native应用的基础知识。但是在开发流程方面还是一头雾水，比如不知道如何构建应用、如何调试应用、如何打包应用等。接下来我们就基于这些内容，以前面用到的BoxApp为主线，整体了解和学习一下构建应用的全过程。

2.3.1 配置文件

这里我们基于iOS上的React Native来创建BoxApp应用。

首先，要确保电脑上已经安装了React Native的命令行工具，如果没有，可以执行下面的代码：

```
npm install -g react-native-cli
```

这样一来命令行工具就安装好了。从代码中可以看到，它用的是Node.js提供的包管理工具NPM来安装的。可见，React Native的命令行工具其实就是Node.js的一个模块。关于Node.js的用法，可以参考https://nodejs.org/en/。安装完命令行工具后，就会有一个全局的 react-native命令可以使用。

接下来，我们使用这个命令行工具来创建BoxApp应用。在开发目录下，运行下面的代码：

```
react-native init Box
```

这段代码的大概执行流程是先获取React Native的源码和依赖包，然后在当前目录下创建一个名叫Box的Xcode项目，这样就创建好了BoxApp应用。如果创建不成功，可能是react-native的地址访问不了啦。如果已经安装过淘宝的cnpm的话，可以通过淘宝的cnpm来安装react-native，此时需要找到文件/usr/local/lib/node_modules/react-native-cli/index.js，将其中的

```
run('npm install --save react-native', function(e) {
```

修改为：

```
run('cnpm install --save react-native', function(e) {
```

当然，如果不想每次都重新获取最新的react-native代码，也可以指定一个在自己电脑上已经安装好的react-native的路径，然后创建一条复制命令替代掉上面的命令。

接下来运行cd Box/命令，进入Box目录，此时会看到如图2-46所示的代码结构。

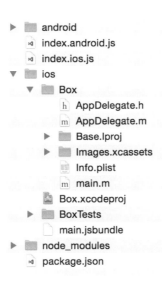

图2-46　Box目录结构

虽然这里有很多文件会被自动创建出来，但是需要修改的只有两个文件——AppDelegate.m 和index.ios.js。下面我们具体介绍一下这两个文件。

- **AppDelegate.m**。这是应用程序入口文件中被调用到的文件。打开该文件，会看到下面的代码：

```
jsCodeLocation = [NSURL URLWithString:@"http://localhost:8081/index.ios.bundle"];
```

这行代码的意思是将jsCodeLocation变量赋值为http://localhost:8081/index.ios.bundle。当应用启动运行的时候，会自动拉取这个bundle文件，该文件里存放的是应用的全部逻辑代码。我们并没有在目录中找到这个文件，那它是怎么来的呢？事实上，这个地址只是一个请求地址，而非真正的静态资源文件地址。那么，这个地址返回的内容是从哪里来的呢。我们带着疑问来继续往下学习。

- **index.ios.js**。这是开发人员编写代码的入口，项目中涉及的业务组件的定义以及组件之间的嵌套都会在这里得到体现。我们在项目的根目录下找到这个文件，用编辑器打开，会看到里面已经默认写了一些代码，这是React Native自动生成的代码架构，是可以正常运行的，有兴趣的同学可以一试，效果图如图2-47所示。

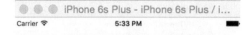

图2-47　React Native项目默认效果

下面我们来分析下这段代码：

```
/**
 * Sample React Native App
 * https://github.com/facebook/react-native
 */
'use strict';
var React = require('react-native');                                    ①
var {
  AppRegistry,
  StyleSheet,
  Text,
  View,
} = React;                                                              ②
var styles = StyleSheet.create({                                        ③
  container: {
  flex: 1,
  justifyContent: 'center',
```

```
    alignItems: 'center',
    backgroundColor: '#F5FCFF',
  },
  welcome: {
    fontSize: 20,
    textAlign: 'center',
    margin: 10,
  },
  instructions: {
    textAlign: 'center',
    color: '#333333',
    marginBottom: 5,
  },
});                                                                  ④
var Box = React.createClass({                                        ⑤
  render: function() {
    return (
      <View style={styles.container}>
        <Text style={styles.welcome}>
          Welcome to React Native!
        </Text>
        <Text style={styles.instructions}>
          To get started, edit index.ios.js
        </Text>
        <Text style={styles.instructions}>
          Press Cmd+R to reload,{'\n'}
          Cmd+D or shake for dev menu
        </Text>
      </View>
    );
  }
});                                                                  ⑥
AppRegistry.registerComponent('Box', () => Box);                     ⑦
```

代码的大体结构如下所示。

(1) 引入React Native包，并导出需要用到的变量，相关代码详见第①~②行。

(2) 创建App需要的样式类styles，相关代码见第③~④行。

(3) 创建App的逻辑组件Box，并使用样式类styles中的样式，相关代码见第⑤~⑥行。

(4) 最后将App组件Box渲染到容器中，相关代码见第⑦行。

接下来我们要做的就是在这个架构的基础上添加BoxApp需要的逻辑代码，这时可以直接复制2.2节的代码过来。至此，我们的代码已经准备就绪，接下来就是让BoxApp运行起来了。

2.3.2 运行

要让BoxApp运行起来，其中涉及编译、打包、启动服务等，大体流程如下所示。

(1) 进入运行目录，相关命令如下：

```
cd Box/iOS
```

(2) 打开文件Box.xcodeproj。双击该文件，便会自动打开Xcode编辑器，然后选择好模拟器类型，这里选择iPhone 6 Plus。

(3) 点击"运行"按钮，也可以使用快捷键⌘+R，便会触发Xcode的编译功能，待Xcode编译成功后，会提示Build Succeeded。若失败，则会提示Build Failed，同时会输出错误日志。

(4) 启动包服务器。

接下来我们会看到，Xcode会自动启动我们的命令行终端，我们把这个终端称为"packager终端"，并在项目的根目录下执行如下命令：

```
npm start
```

这个命令用于执行Node.js启动服务，启动成功后就会在终端输出成功的日志：

```
Running packager on port 8081.

Keep this packager running while developing on any JS projects. Feel
free to close this tab and run your own packager instance if you
prefer.

https://github.com/facebook/react-native
```

```
Looking for JS files in
  /Users/username/Developer/ReactNative/Box

React packager ready.
```

当然，也经常会遇到启动失败的情况，常见的原因是端口被占用了。这有可能是上个终端启动后没有关闭导致的，所以在用Xcode编译应用前，尽量将所有终端都关闭，以免带来不必要的麻烦。在应用编译的过程中，Xcode会自动开启需要的终端，也就是上面的"packager终端"。注意这个终端在应用的调试中要一直保持开着，直到Xcode需要重新编译时再把它关掉。我们上面讲到的http://localhost:8081/index.ios.bundle之所以能够访问，靠的就是这个服务通过动态分析index.io.js中的依赖，并对其进行合并得到的。而且该服务还有个好处就是允许代码实时渲染。这样我们修改了JavaScript文件的代码，就不需要重新通过Xcode编译也能看到效果。

(5) 启动模拟器。

启动包服务器后，Xcode便会自动启动我们配置好的模拟器，待模拟器成功启动后，就可以看到BoxApp的页面了。修改代码之后，只需要在模拟器上按快捷键⌘+R即可看到最新的效果了。有时候我们觉得模拟器太大啦，此时也可以通过选择Simulator→Window→Scale来调整模拟器窗口的大小。

(6) 修改端口。

包服务器默认的启动端口为8081，如果与现有的服务器端口有冲突的话，可以将其改为自己想要的端口。不过需要注意的是，要同时修改下面几个地方。

- **包服务器启动时监听的端口**。在文件Box/node_modules/react-native/packager/packager.js中找到下面的代码：

```
var options = parseCommandLine([{
  command: 'port',
  default: 8081,
  type: 'string',
},
......
```

然后将8081改为需要的端口。

- **应用程序启动时，读取所有逻辑代码的地址**。在文件Box/ios/Box/AppDelegate.m中找到下面的代码：

```
jsCodeLocation = [NSURL URLWithString:@"http://localhost:8081/index.ios.bundle"];
```

然后将8081修改为需要的端口。

- **WebSocket监听的端口**。在文件Box/node_modules/react-native/Libraries/WebSocket/RCTWebSocketExecutor.m中找到下面的代码：

```
- (instancetype)init
{
  return [self initWithURL:[RCTConvert NSURL:@"http://localhost:8081/debugger-proxy"]];
}
```

然后将8081改为需要的端口。

2.3.3 调试

在上面的模拟器中，我们的逻辑JavaScript文件是服务端包服务器动态编译出来的，不是实际存在的静态文件。当需要调试的时候，该怎么办呢？在模拟器上按快捷键⌘+D，就会弹出如图2-48所示的调试菜单。

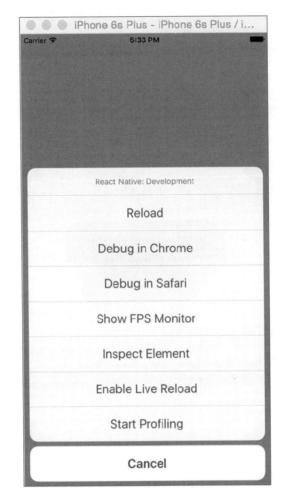

图2-48 React Native Debug界面

下面我们逐一了解这些调试功能。

- Reload。相当于刷新页面，快捷键为⌘+R。需要注意的是，如果你新增了文件或者修改了Native代码，需要使用Xcode重新编译应用，而不是刷新页面。只有修改的是JavaScript文件时，刷新功能才起效果。
- Debug in Chrome。该功能允许开发人员在Chrome中调试应用，其调试方式和调试Web应用一样。当该功能被点击时，React Native会启动Chrome浏览器，并且打开一个http://localhost:8081/debugger-ui的新标签。在这个标签中，打开开发者工具，选择Console，就可以看到输出的日志信息，如图2-49所示。

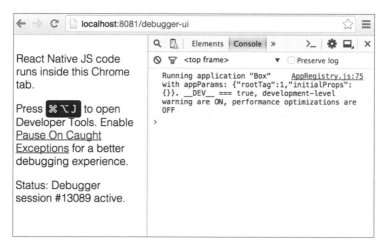

图2-49　Console调试界面

这里我们也可以调试JavaScript代码。打开Sources，选择localhost:8081下的index.ios.bundle文件并点击，此时在右侧便可以看到动态合并好的代码，如图2-50所示。当然，也可以给它添加断点来调试。

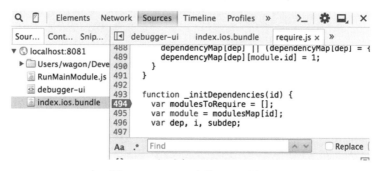

图2-50　Sources中的Debug界面

- Debug in Safari。该功能同Debug in Chrome类似，这里就不赘述了。
- Live Reload。该功能主要用来实现自动刷新，当我们点击Enable Live Reload时，如果应用中的JavaScript代码有任何修改，它都会自动帮我们更新，不需要人为去操作刷新功能。当然，也可以通过点击Disable Live Reload来禁用该功能。
- Show FPS Monitor。该功能是用来对UI和JavaScript的FPS的监控，其效果如图2-51所示。

图2-51　FPS Monitor界面

说明　FPS是图像领域中的定义,是指画面每秒传输的帧数,通俗来讲就是指动画或视频的画面数。FPS是测量用于保存、显示动态视频的信息数量。每秒钟的帧数愈多,所显示的动作就会愈流畅。通常,要避免动作不流畅的最低帧数是30。某些计算机视频格式,每秒只能提供15帧。

- Inspect Element。做过Web开发的人员应该都用过Chrome的查找元素功能,它可以帮我们很方便地发现当前元素的位置、样式、层级关系等,是Web开发中不可缺少的一个功能。这里我真佩服Facebook的强大,现在也为React Native提供了该功能,而且还提供了监控应用性能的功能。

对应用性能的监控如图2-52所示。

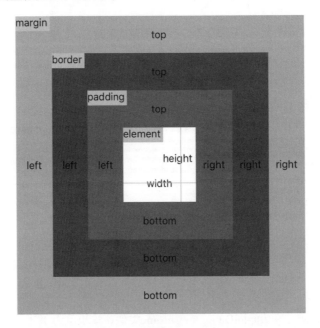

图2-52　应用性能监控界面

对应用元素Box Model的监控如图2-53所示。

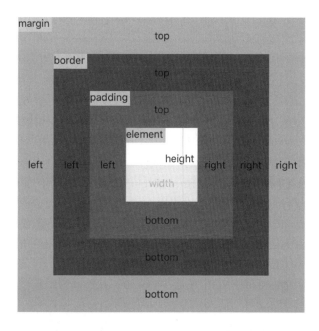

图2-53 元素盒子模型检查界面

在图2-53中,我们看到当前选中的元素颜色变反选,而且在该元素的下边列出了padding、margin、width、height等盒子模型的详细信息,便于我们快速定位问题。

- **Profiling**。该功能主要用来监控应用在一段时间内的指标信息。既然是一段时间,就要有一个开始动作和结束动作。当我们要监控应用指标时,需要在操作前选择Start Profiling,然后在操作结束后选择Stop Profiling,这样就会弹出一个提示框,如图2-54所示。

它提示我们数据已经生成,同时会在包服务端看到下面的日志信息:

```
NOTE: Your profile data was kept at:

  /tmp/dump_1443827124505.json
```

图2-54 指标监控成功后的弹出界面

打开这个JSON文件,就可以看到应用在这段时间内的详细指标信息了。如果我们装了Google trace-viewer插件,包服务器会帮我们自动调用trace2html /tmp/dump_1443827124505.json命令打开该JSON文件。

此外,还要介绍一个重要的调试工具react-developer-tools,它在图2-44中并没有涉及,但在日常开发中经常用到,所以这里也单独拿出来讲一下。它是一个Chrome浏览器的扩展,能详细显示出元素的层级关系,以及每个组件对应的state、props等属性的详细信息。它的安装地址为https://chrome.google.com/webstore/detail/react-developer-tools/fmkadmapgofadopljbjfkapdkoienihi?hl=en。它是配合Debug in Chrome中打开控制台使用的,如果控制台打开,你会看到最后有个react标签,点击这个标签,可能会看到一个正在连接的提示,此时回到模拟器点击一下,就会看到当前应用的结构啦,如图2-55所示。

图2-55 React的Chrome插件界面

这里可以很方便地看到各组件之间的嵌套关系以及每个组件的事件、属性、状态等信息。

前面讲到的都是基于模拟器的调试,那么真机上的调试是怎样的呢?这里阐述下真机调试的流程。

(1) 配置好签名证书,并将打包环境设置成Debug模式。

(2) 将2.3.2节中"(6)修改端口"涉及的localhost改为对应的IP,而且需要调试的真机和该IP保持在一个网段。将调试的真机插在电脑上,在Xcode中选择这个真机作为Simulator。

(3) 运行编译命令,这样待编译完成后就会在真机设备上启动应用了。要在真机上要调出调试菜单,只需要摇一摇手机即可。但前提是一定要把打包环境设置为Debug模式,否则调试菜单出不来。菜单出来之后的调试方法就和模拟器一样了，这里也就不再细说啦。

2.3.4 内部发布

这里把它叫做内部发布而不叫发布，是因为生成的是BoxApp的包，没有上App Store，但内部人员可以安装和使用。

前面我们讲的都是电脑上模拟器的处理，这节中主要讲离线状态下手机设备的使用情况。

还是进入到上面讲到的AppDelegate.m中，将之前的

```
jsCodeLocation = [NSURL URLWithString:@"http://localhost:8081/index.ios.bundle"];
```

注释掉，并且将

```
//jsCodeLocation = [[NSBundle mainBundle] URLForResource:@"main" withExtension:@"jsbundle"];
```

这句的注释去掉。

接下来，在项目跟目录下运行

```
curl http://localhost:8081/index.ios.bundle -o main.jsbundle
```

或者运行

```
react-native bundle
```

此时就会在项目和目录下生成main.jsbundle文件，这个文件将所有编写的JavaScript文件都打包在一起了。

然后配置好自己的证书，再通过Xcode的编译运行，就可以得到一个离线的ipa包，我们装在真机上就可以在无网络的情况下运行BoxApp了。

2.4 参考资料

本章的参考资料如下：

- http://facebook.github.io/react-native/docs/getting-started.html
- https://github.com/facebook/react-native
- http://npm.taobao.org/
- https://nodejs.org/en/
- http://facebook.github.io/react/docs/getting-started.html
- http://calendar.perfplanet.com/2013/diff/
- http://www.w3.org/TR/css-flexbox/
- https://css-tricks.com/snippets/css/a-guide-to-flexbox/
- http://facebook.github.io/react-native/docs/flexbox.html#content

第二部分
API 和组件篇

- 第3章　常用组件及其实践
- 第4章　常用API及其实践
- 第5章　Native扩展
- 第6章　组件封装

第3章 常用组件及其实践

通过第1章和第2章,我们基本了解了React Native的开发环境及其相关的基础知识。那么,在这一章中,我们要着重介绍React Native的相关组件。只有掌握了React Native的组件,才能熟练开发App。当然,React Native的魅力在于能够使用iOS的原生组件和原生API。因此,学习和掌握React Native组件和API是十分必要的。

3.1 View 组件

就像学习HTML一样,标签十分重要。开发Web应用程序时,需要使用很多的HTML标签,例如form、h1、canvas等。但是在基于DIV+CSS布局的Web开发中,最为重要的一个元素就是div。因为div是页面布局的基础,是作为容器元素存在的。在React Native中,就有一个类似于div的组件,那就是View组件。

3.1.1 View 介绍

View是一个容器组件,提供了视图布局的功能,是UI组件中最基本的组件。它可以多层嵌套,支持flexbox布局,起到容器组件的作用。使用View组件,可以进行复杂的布局和精巧的页面设计。

3.1.2 案例:九宫格实现

对View组件有了个初步的认识后,现在我们来使用View组件做一些有趣的事情——将View组件用于实战中。这里,我们使用携程的客户端首页举例,其Web App的地址是http://m.ctrip.com。访问该地址,可以看到其首页如图3-1所示。

那么,我们需要实现的功能是中间的"九宫格"布局。如图3-1所示,"酒店"一栏布局的方式有很多种。

图3-1 携程首页

水平3栏，第二栏和第三栏分别上下两栏，如图3-2所示。

1 酒店	海外酒店 2 特惠酒店	团购 3 客栈.公寓

图3-2 第一种布局方式

水平两栏，第二栏分为上下两栏，再进一步将上下两栏分为左右两栏，如图3-3所示。

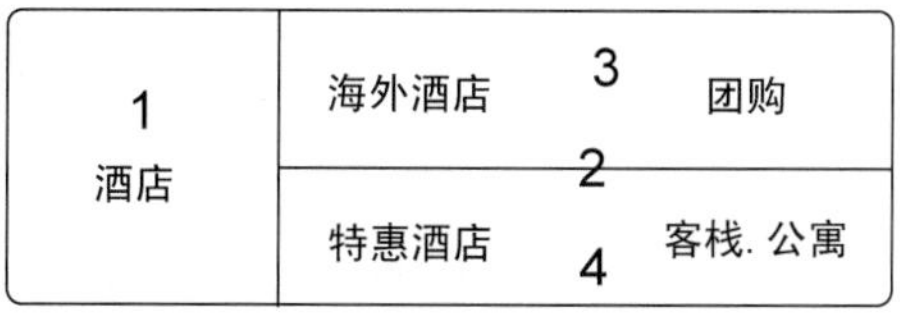

1 酒店	海外酒店　　3　　团购 2 特惠酒店　　4　　客栈.公寓

图3-3 第二种布局方式

这里我们使用第一种方式进行布局。现在按照如下步骤实现携程App的"九宫格"。

(1) **加载View组件**。首先，创建项目，然后在index.ios.js中引入react-native并加载View组件，代码如下所示：

```
var React = require('react-native');
var {
  AppRegistry,
  StyleSheet,
  View
} = React;
```

其中AppRegistry和StyleSheet是React Native提供的API，前者负责注册App入口组件，后者负责创建样式表。

(2) **创建组件**。我们已经引入了React Native，因此使用React.createClass创建一个组件，即app，它将作为应用程序的入口组件。相关代码如下：

```
var app = React.createClass({
  render: function(){

  }
});
```

在React.createClass中，需要有一个render方法，该方法负责渲染视图。同时，render方法需要返回一个JSX对象（包含null），并且该对象只能包含在一个节点中。也就是说，返回的对象必须有且只有一个容器对象包裹。当然，我们会在后续的步骤中补全render方法。

(3) **添加样式表**。当然，我们也可以使用内联样式。这里建议在View组件上使用外部样式，而不是内联样式，因为这样更加容易维护和更新样式。下面创建一个样式对象styles，具体的代码如下所示：

```
var styles = StyleSheet.create({

});
```

在上述代码中，StyleSheet.create用于创建一个样式对象，其中传入的参数是一个JavaScript字面量对象。这里建议一个组件使用一个StyleSheet对象，而不是多个组件共用一个样式对象，因为这样组件的功能更加细化和解耦，更能体现组件的封装，也利于后续开发者的程序维护。

(4) **注册入口**。这个应用程序应该有一个入口，这样就能根据入口组件动态加载其他组件。React Native提供了AppRegistry API来做这件事，相关代码如下所示：

```
AppRegistry.registerComponent('APP', () => app);
```

registerComponent方法的第一个参数是我们应用程序的名称，即项目的名称，第二个参数是入口组件对象，即第(2)步创建的app对象。在4.1节中，我们会详细介绍AppRegistry。当然，现在你也可以翻阅到4.1节查看。

(5) **外层布局**。这里我们选择了第一种方式,即水平3栏,然后第2栏和第3栏分别上下两栏,如图3-2所示。首先在View中嵌套3个View组件,代码如下所示:

```
var app = React.createClass({
  render: function(){
    return (
      <View>
       <View></View>
       <View></View>
       <View></View>
      </View>
    );
  }
});
```

我们清楚地看到这3个View组件外层包裹了一个View组件,也就是第(2)步所说的"有且只有一个容器对象包裹"。其实,这个时候使用快捷键cmd+R是看不到效果的,因为视图渲染中既没有边框也没有文字。下一步,我们让它可视化。

(6) **flexbox水平三栏布局**。我们需要将3个View组件水平布局并且同时平分屏幕宽度。React Native支持flexbox布局,2.3节已经详细介绍过。具体的代码如下:

```
var app = React.createClass({
  render: function(){
    return (
      <View style={styles.container}>
        <View style={styles.item}></View>
        <View style={styles.item}></View>
        <View style={styles.item}></View>
      </View>
    );
  }
});

var styles = StyleSheet.create({
  container:{
    flex:1,
    borderWidth:1,
    borderColor: 'red',
    flexDirection: 'row'
  },
  item:{
    flex:1,
    height:80,
    borderColor:'blue',
    borderWidth:1,
  }
});
```

这里我们使用了StyleSheet创建样式对象,共创建了两个样式:container和item。container作用于最外层的View组件上,而flex:1的意思是将最外层的View组件平铺占满整个屏幕,并且边框的宽度是1pt（其实，你也可以将其想象成Web开发中的px，即像素），边框的颜色是红色。这里有个比较重要的样式，即flexDirection。因为在React Native中，flexbox默认组件的布局方式是纵向布局，即flexDirection的默认值是column。因此，为了将3个View组件水平布局，需要将纵向布局调整为水平布局，即flexDirection的值设为row。这里带有container样式的组件都按照水平方式布局。item样式作用于内部的3个View组件。我们设定内部3个View组件的高度为80，边框的宽度是1，边框的颜色是蓝色。我们注意到item样式上的flex的值也是1，这是因为我们需要将3个View组件平分屏幕宽度。flexbox的flex属性代表的是权重，内部3个View组件的flex值都是1，所以3个View分别占1/3。那么，这个1/3具体指什么呢？因为我们在父层节点上指出了按照水平布局，即flexDirection: 'row'，所以这3个View组件平分的就是父元素的宽度。在这个例子中，是三等分了屏幕的宽度。

同时，我们也注意到了View组件有个重要的属性——style，该属性是所有组件都支持的，其属性值是{JavaScript JSON对象}的形式。在上面的例子中，我们看到style={styles.container}，这就是引用了外部样式。如果需要写成内联样式，则应该像如下代码所示：

```
style={{flex:1, borderWidth:1, borderColor: 'red', flexDirection: 'row'}}
```

如果需要包裹多个样式类，则使用如下代码：

```
style={[styles.style1, styles.style2]}
```

同时包裹样式类或者内联样式如下列代码所示：

```
style={[styles.style1, {flex:1, borderWidth:1}]}
```

当然，我们不建议过多使用这种方式。

(7) **上下两栏布局**。我们看到"海外酒店"和"特惠酒店"是在第2栏上下布局，"团购"和"客栈.公寓"是在第3栏上下布局。因此，我们在第2栏和第3栏中分别嵌入两个View组件。为了方便显示，这里引入了Text组件，相关代码如下：

```
var {
  AppRegistry,
  StyleSheet,
  View,
  Text
} = React;
```

现在就可以在View组件里面使用Text组件了，这样文本就会显示在界面中。3.2节会对Text组件做详细介绍，感兴趣的读者可以翻阅查看。因为View组件默认是纵向（垂直）布局，所以这里只需要给上下两个View组件加上flex:1的样式就可以了，这样"海外酒店"和"特惠酒店"就是垂直平分外层View组件高度。具体的代码如下：

```
var app = React.createClass({
  render: function(){
    return (
      <View style={styles.container}>
```

```
        <View style={[styles.item, styles.center]}>
          <Text>酒店</Text>
        </View>
        <View style={styles.item}>
          <View style={[styles.center, styles.flex]}>
            <Text>海外酒店</Text>
          </View>
          <View style={[styles.center, styles.flex]}>
            <Text>特惠酒店</Text>
          </View>
        </View>
        <View style={styles.item}>
          <View style={[styles.center, styles.flex]}>
            <Text>团购</Text>
          </View>
          <View style={[styles.center, styles.flex]}>
            <Text>客栈.公寓</Text>
          </View>
        </View>
      </View>
    );
  }
});

var styles = StyleSheet.create({
  container:{
    flex:1,
    borderWidth:1,
    borderColor: 'red',
    flexDirection: 'row'
  },
  item:{
    flex:1,
    height:80,
    borderColor:'blue',
    borderWidth:1,
  },
  center:{
    justifyContent:'center',  /*垂直居中，实际上是按照flexDirection的方向居中*/
    alignItems: 'center'      /*水平居中*/
  },
  flex:{
    flex: 1
  }
});
```

到目前为止，我们完成的效果还比较粗糙，如图3-4所示。

图3-4 初步效果

(8) 完善效果。其实，这样的效果是达不到产品要求的，我们需要进一步完善样式，使其更加符合我们的预期。这里我们进行了如下处理：背景颜色、线条的分割、字体的设置、圆角处理等。具体的完成代码如下所示：

```javascript
var React = require('react-native');

var {
  AppRegistry,
  StyleSheet,
  View,
  PixelRatio,
  Text
} = React;

var app = React.createClass({
  render: function(){
    return (
      <View style={styles.flex}>
        <View style={styles.container}>
          <View style={[styles.item, styles.center]}>
            <Text style={styles.font}>酒店</Text>
          </View>
          <View style={[styles.item, styles.lineLeftRight]}>
            <View style={[styles.center, styles.flex, styles.lineCenter]}>
              <Text style={styles.font}>海外酒店</Text>
            </View>
            <View style={[styles.center, styles.flex]}>
              <Text style={styles.font}>特惠酒店</Text>
            </View>
          </View>
          <View style={styles.item}>
            <View style={[styles.center, styles.flex, styles.lineCenter]}>
              <Text style={styles.font}>团购</Text>
            </View>
            <View style={[styles.center, styles.flex]}>
              <Text style={styles.font}>客栈.公寓</Text>
            </View>
          </View>
        </View>
      </View>
    );
  }
});

var styles = StyleSheet.create({
  container:{
    marginTop:25,
    marginLeft:5,
    marginRight:5,
    height:84,
    flexDirection: 'row',
    borderRadius:5,
```

```
    padding: 2,
    backgroundColor:'#FF0067',
  },
  item:{
    flex:1,
    height:80,
  },
  center:{
    justifyContent:'center',  /*垂直居中,实际上是按照flexDirection的方向居中*/
    alignItems: 'center'      /*水平居中*/
  },
  flex:{
    flex: 1
  },
  font:{
    color:'#fff',
    fontSize:16,
    fontWeight:'bold'
  },
  lineLeftRight:{
    borderLeftWidth:1/PixelRatio.get(),
    borderRightWidth:1/PixelRatio.get(),
    borderColor: '#fff'
  },
  lineCenter:{
    borderBottomWidth:1/PixelRatio.get(),
    borderColor: '#fff'
  }
});

AppRegistry.registerComponent('APP', () => app);
```

完善后的界面效果如图3-5所示。

图3-5 简单的"九宫格"效果

在前面的代码中,我们做了细微的调整,具体如下。

- 引入了 PixelRatio API,PixelRatio的get方法用于获取高清设备的像素比。使用 1/PixelRatio.get()就可以获得最小线宽。
- container使用了margin属性,marginTop: 25使得内容距离状态栏25pt,marginLeft:5和 marginRight:5分别用于设置距离屏幕左边和右边5pt。
- 设置字体为16pt、白色、粗体。

3.2 Text 组件

Text组件主要用于显示文本。当然，它具有响应特性，该特性表现为被触摸时是否高亮。它同样支持多层嵌套，所以可以继承样式。在Web开发中，字体样式的继承十分重要。比如网页大部分地方的字体都是12px，此时我们可以在body上设置所有字体是12px。但是React Native是不支持这种继承的，字体样式只有在Text组件上才起作用。因此，字体样式的继承也只能通过Text组件来实现。内部的Text组件可以继承外部Text组件的样式。

3.2.1 Text 组件介绍

很多时候，我们不期待一个Text组件具有过多的功能，而仅仅是显示文本。React Native的设计也是如此，但是Text组件还是有许多特性需要我们注意的。Text组件的重要性不言而喻，无论是Web开发还是客户端开发，都离不开Text组件。Text组件常用的特性如下所示。

- **onPress**：该属性的值是一个函数，支持按下事件（即手指触摸事件）。当手指按下的时候，执行该函数。
- **numberOfLines**：该属性的值是一个数字，用于规定最多显示多少行，如果超过该数值，则以省略号（...）表示。
- **onLayout**：该属性的值是一个函数，用于获取该元素布局的位置和大小，例如：{"target":4,"layout":{"y":10,"width":300,"x":10,"height":117}}。一般事件函数的形式是function(e){consolc.log(e.nativeEvent)};，这样就可以打印事件的参数。

3.2.2 案例：网易新闻列表展示

3.2.1节中介绍的Text组件的属性并不多，但是，它的有些特性在实践中还是需注意的。我们不仅需要做理论的拜读者，还要在企业开发实践中去验证理论和踩坑。在这一节中，我们需要实践的是网易新闻的一个页面，如图3-6所示。

图3-6 网易新闻

首先，我们来分析该页面的结构。这个页面的布局相来说比较简单，分为上、中、下三栏布局。React Native提倡组件化，那么组件化的颗粒度为多少主要取决于应用的结构设计。这里，上面的头部可以是一个组件，中间的文章标题列表可以是一个组件，下面的"今日要闻"可以是一个组件。我们不希望该页面过于复杂，不希望后期的维护成本很高。因此，我们将头部组件独立成一个文件，而将列表组件和"今日要闻"作为独立组件放在同一个文件内。这样做的好处是头部组件可以被其他页面共用，列表组件和"今日要闻"组件只能在当前页面使用。我们规划的结构如图3-7所示。

图3-7 网易新闻页面结构图

下面我们一步步来实现这个页面。

1. 封装头部组件

打开网易的Web App（http://news.163.com/mobile/），可以发现网易的头部标题其实是一张SVG格式的图片：http://img1.cache.netease.com/f2e/news/index2015/img/logo.svg。这里为了更好地练习Text组件，我们使用该组件实现类似的效果。我们将"网易新闻有态度°"分为3个Text组件来实现，这样就可以表达3种不同的效果，相关代码如下所示：

```
var React = require('react-native');

var {
  AppRegistry,
  StyleSheet,
  View,
  Text
} = React;

var Header = React.createClass({
  render: function(){
```

```
      return (
        <View style={styles.flex}>
          <Text style={styles.font}>
            <Text style={styles.font_1}>网易</Text>
            <Text style={styles.font_2}>新闻</Text>
            <Text>有态度°</Text>
          </Text>
        </View>
      );
    }
});

var styles = StyleSheet.create({
  flex:{
    marginTop:25,
    height:50,
    borderBottomWidth:3/React.PixelRatio.get(),
    borderBottomColor:'#EF2D36',
    alignItems:'center' /* 使Text组件水平居中*/
  },
  font:{
    fontSize:25,
    fontWeight: 'bold',
    textAlign:'center' /*使文字在Text组件中居中*/
  },
  font_1:{
    color:'#CD1D1C'
  },
  font_2:{
    color:'#FFF',
    backgroundColor:'#CD1D1C',
  }
});

module.exports = Header;
```

这里我们将头部封装成了一个简单的组件，在代码最后我们将其export成独立的模块。在上面的代码中我们发现，Text组件嵌套之后就不会按照flexbox布局。当然，这也符合我们的初衷，我们本来希望Text组件用来做一些文本的展示。我们在最外层的Text组件上定义了font样式，即规定了该Text组件内部的所有Text组件的字体是25pt，字体加粗并居中显示。同时为了对第一个和第二个Text组件做一些特别的处理，我们给它们分别加上了font_1和font_2样式。这里font_1的字体是红色，font_2的字体背景是红色，字体是白色。为了测试组件是否可用，我们将以上代码放入header.js文件中，然后在index.ios.js中加载header.js文件中的Header组件，具体的代码如下：

```
var React = require('react-native');
var Header = require('./header');

var {
  AppRegistry,
  StyleSheet,
  View,
```

```
    Text
} = React;

var app = React.createClass({
  render: function(){
    return (
      <View style={styles.flex}>
        <Header></Header>
      </View>
    );
  }
});

var styles = StyleSheet.create({
  flex:{
    flex:1
  },

});

AppRegistry.registerComponent('APP', () => app);
```

在上述代码中，var Header = require('./header')这句代码表明我们使用require函数加载了Header组件。这里我们使用组件的方式也很简单：<Header></Header>。头部组件运行的效果如图3-8所示。

图3-8 头部组件效果

2. 列表组件

这里我们希望将新闻标题做成列表，而每一条新闻标题实际上可以独立成一个简单的组件。同时也希望将标题的数据传入到组件中，而不是写死在组件上。具体的代码如下：

```
var List = React.createClass({
  render: function(){
    return (
      <View style={styles.list_item}>
        <Text style={styles.list_item_font}>{this.props.title}</Text>
      </View>
    );
  }
});
var styles = StyleSheet.create({
  flex:{
    flex:1
  },
```

```
    list_item:{
      height:40,
      marginLeft:10,
      marginRight:10,
      borderBottomWidth:1,
      borderBottomColor: '#ddd',
      justifyContent: 'center'
    },
    list_item_font:{
      fontSize:16
    }
});
```

这里我们使用React.createClass创建了List组件，同时使用this.props属性来接收外部传入的参数。使用List组件的方法如下：

```
var app = React.createClass({
  render: function(){
    return (
      <View style={styles.flex}>
        <Header></Header>
        <List title='宇航员在太空宣布"三体"获奖'></List>
        <List title='NASA发短片纪念火星征程50年'></List>
        <List title='男生连续做一周苦瓜吃吐女友'></List>
        <List title='女童遭鲨鱼袭击又下海救伙伴'></List>
      </View>
    );
  }
});
```

这里我们只是在List组件上加入title属性即可。此时列表组件其实已经可以投入生产了，具体的效果如图3-9所示。

图3-9　新闻列表

3. 完成"今日要闻"

我们在"今日要闻"组件上要完成3个功能：

❑ 文章标题的展示；

❑ 文章标题超过两行时用省略号（...）代替；

❑ 点击标题弹出标题内容。

我们将整块功能封装成一个组件，具体的实现代码如下：

```
var ImportantNews = React.createClass({
  show: function(title){
    alert(title);
  },
  render: function(){
    var news = [];
    for(var i in this.props.news){
      var text = (
        <Text
          onPress={this.show.bind(this, this.props.news[i])}
          numberOfLines={2}
          style={styles.news_item}>
          {this.props.news[i]}
        </Text>
      );
      news.push(text);
    }
    return (
      <View style={styles.flex}>
        <Text style={styles.news_title}>今日要闻</Text>
        {news}
      </View>
    );
  }
});
var styles = StyleSheet.create({
  flex:{
    flex:1
  },
  list_item:{
    height:40,
    marginLeft:10,
    marginRight:10,
    borderBottomWidth:1,
    borderBottomColor: '#ddd',
    justifyContent: 'center'
  },
  list_item_font:{
    fontSize:16
  },
  news_title:{
    fontSize:20,
    fontWeight:'bold',
    color: '#CD1D1C',
    marginLeft:10,
    marginTop:15,
  },
  news_item:{
    marginLeft:10,
```

```
    marginRight:10,
    fontSize:15,
    lineHeight:20,
  }
});
```

在上面的代码中,我们给Text组件增加了onPress事件。这里,当新闻被按下的时候,会弹出新闻的标题。需要注意的是,这里传递参数时需要使用bind方法,该方法的第一个参数是上下文对象,第二个参数是传递参数;同时需要使用numberOfLines指定标题最多两行。此外,我们使用this.props获取传递的数据。ImportantNews组件默认只需要传递一个新闻的数组即可,这里使用了alert函数,该函数是全局的,可以调用系统的弹出信息窗。调用ImportantNews组件时,只需一句代码即可,具体如下所示:

```
<ImportantNews news={[
  '1、刘慈欣《三体》获"雨果奖"为中国作家首次',
  '2、京津冀协同发展定位明确:北京无经济中心表述',
  '3、好奇宝宝第一次淋雨,父亲用镜头记录了下来',
  '4、人民邮电出版社即将出版《React Native入门与实战》,读者可以使用JavaScript开发原生应用']}>
</ImportantNews>
```

4. 最终效果和代码

现在我们已经完成了大部分的内容,得到的效果如图3-10所示。

图3-10　网易新闻效果

该节的完整代码可参考https://github.com/vczero/React-Native-Code项目中的第3章3.2节。

3.3 NavigatorIOS 组件

应用程序往往由很多功能视图组成。就像Web App一样，多页应用自然是多个页面，单页应用也会存在不同路由。因此，一个应用中最为重要的功能之一就是"路由"（或者说是"导航"）。因为只有存在路由，才能实现视图之间的切换和前进、后退。在React Native中，就存在一个专门负责视图切换的组件，它就是NavigatorIOS组件。该组件具有很多有用的方法和属性，可以很方便地让我们进行路由管理。当然，React Native也提供了一个兄弟组件，那就是Navigator。如果需要了解Navigator的用法，可以翻阅到10.3.3节查看。

3.3.1 NavigatorIOS 组件介绍

该组件本质上是对UIKit navigation的包装。也就是说，使用NavigatorIOS进行路由切换，实质上是调用了UIKit的navigation。

路由是一个JavaScript对象，代表着一个页面（或者说视图）组件。NavigatorIOS组件默认的路由提供了initialRoute属性。示例代码如下：

```
render: function() {
  return (
    <NavigatorIOS
      initialRoute={{
        component: MyView,
        title: 'My View Title',
        passProps: { myProp: 'foo' },
      }}
    />
  );
}
```

在上面的代码中，component表示该页面需要加载的组件视图，title表示需要在头部显示的标题，passProps用于页面间传递数据。

这里简要介绍一下NavigatorIOS组件的属性，具体如下所示。

- **barTintColor**：导航条的背景颜色。
- **initialRoute**：初始化路由。路由对象如下所示：

```
{
  component: function, //加载的视图组件
  title: string, //当前视图的标题
  passProps: object, //传递的数据
  backButtonIcon: Image.propTypes.source,//后退按钮图标
  backButtonTitle: string,//后退按钮标题
  leftButtonIcon: Image.propTypes.source, //左边按钮图标
```

```
leftButtonTitle: string, //左边按钮标题
onLeftButtonPress: function,//左边按钮点击事件
rightButtonIcon: Image.propTypes.source, //右边按钮图标
rightButtonTitle: string, //右边按钮标题
onRightButtonPress: function, //右边按钮点击事件
wrapperStyle: [object Object]//包裹样式
}
```

- **itemWrapperStyle**：为每一项定制样式，例如设置每个页面的背景颜色。
- **navigationBarHidden**：当其值为true时，隐藏导航栏。
- **shadowHidden**：是否隐藏阴影，其值为true或者false。
- **tintColor**：导航栏上按钮的颜色设置。
- **titleTextColor**：导航栏上字体的颜色。
- **translucent**：导航栏是否是半透明的，其值为true或者false。

在组件视图切换的时候，navigator会作为一个属性对象被传递。我们可以通过this.props.navigator获得navigator对象。navigator是一个十分重要的对象，它可以控制路由的跳转和组件的加载。因此，要掌握NavigatorIOS组件，必须掌握navigator对象。navigator对象的主要方法如下所示。

- **push(route)**：加载一个新的页面（视图或者路由）并且路由到该页面。
- **pop()**：返回到上一个页面。
- **popN(n)**：一次性返回N个页面，当N=1时，即相当于pop()方法的效果。
- **replace(route)**：替换当前的路由。
- **replacePrevious(route)**：替换前一个页面的视图并且回退过去。
- **resetTo(route)**：取代最顶层的路由并且回退过去。
- **popToTop()**：回到最上层视图。

可以看到，navigator提供的方法已经很强大了。

3.3.2 案例：列表页跳转详情页

导航栏几乎在所有的应用中都存在，是App页面（视图）跳转的基石。这里我们演示一个简单的例子：从列表页跳转到详情页，如图3-11所示。

为了实现图3-11所示的跳转，需要如下3个组件：

- App入口组件；
- 列表页组件；
- 详情页组件。

我们按照以下步骤完成这个跳转。

3.3 NavigatorIOS 组件

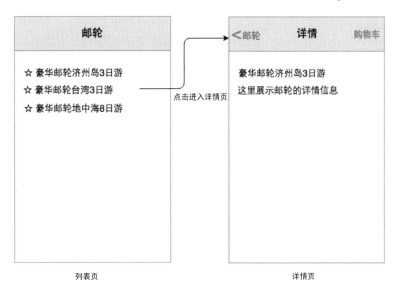

图3-11　原型设计

1. 入口组件

首先，加载NavigatorIOS组件，并将其作为路由跳转的入口，具体代码如下：

```
var NV = React.createClass({
  render: function(){
    return(
      <NavigatorIOS
        style={{flex:1}}
        initialRoute={{
          component: List,
          title: '邮轮',
          passProps: {},
        }}
      />
    );
  }
});
```

这里我们为NavigatorIOS组件配置了一个初始化路由（initialRoute），即List。这样页面启动的时候，就会加载List组件。

2. 列表页组件

这里我们使用ScrollView组件和Text组件构建最简单的列表页组件。为了演示需要，这里展示3条邮轮的信息，具体代码如下：

```
var List = React.createClass({
  render: function(){
    return (
```

```
      <ScrollView style={styles.flex}>
        <Text style={styles.list_item} onPress={this.goTo}>☆ 豪华邮轮济州岛3日游</Text>
        <Text style={styles.list_item} onPress={this.goTo}>☆ 豪华邮轮台湾3日游</Text>
        <Text style={styles.list_item} onPress={this.goTo}>☆ 豪华邮轮地中海8日游</Text>
      </ScrollView>
    );
  },
  goTo: function(){
    this.props.navigator.push({
      component: Detail,
      title: '邮轮详情',
      rightButtonTitle: '购物车',
      onRightButtonPress: function(){
        alert('进入我的购物车');
      }
    });
  }
});
```

在上面的代码中,我们定义了一个List组件。该组件展示了3条邮轮的信息,并且每条信息绑定了一个点击事件。当用户点击某一条邮轮信息时,我们使用this.props.navigator向里面添加Detail视图;并且,Detail视图的标题是"邮轮详情",右边按钮的标题是"购物车"。当点击"购物车"按钮时,会弹出"进入我的购物车"信息框。

3. 详情页组件

在List组件中,我们使用了详情页组件(Detail)。因此,这里需要开发一个简单的详情页组件,具体的代码如下:

```
var Detail = React.createClass({
  render: function(){
    return (
      <ScrollView>
        <Text>详情页</Text>
        <Text>尽管信息很少,但这就是详情页</Text>
      </ScrollView>
    );
  }
});
```

详情页组件很简单,这里只是简单展示了两行文本。

4. 整体代码和效果

现在,我们清楚地知道NV组件调用List组件,List组件调用Detail组件,它们之间形成链式关系,所有的路由都被navigator.push到一个路由数组中,navigator对象对路由进行控制和跳转。该实例的完整代码可以参考https://github.com/vczero/React-Native-Code中的3.3节。项目运行的效果如图3-12所示。

图3-12　NavigatorIOS效果图

3.4　TextInput 组件

在一个应用程序中，输入框是必不可少的，比如"搜索"功能是大部分应用程序都拥有的。TextInput是可以通过键盘将文本输入到App的组件，它提供了比较丰富的功能，例如自动校验、占位符以及指定弹出不同的键盘类型等。

3.4.1　TextInput 组件介绍

我们期待TextInput能帮助我们做更多的事情，而不是我们去模拟一些事件和属性。React Native在TextInput做的还是很好的，属性和事件基本够用。我们既可以用TextInput组件做基本的组件，也可以用TextInput组件做自动补全的搜索功能。TextInput的主要属性和事件如下所示。

- autoCapitalize：枚举类型，可选值有'none'、'sentences'、'words'、'characters'。当用户输入时，用于提示。
- placeholder：占位符，在输入前显示的文本内容。
- value：文本输入框的默认值。
- placeholderTextColor：占位符文本的颜色。
- password：如果为true，则是密码输入框，文本显示为"*"。
- multiline：如果为true，则是多行输入。

- **editable**：如果为false，文本框不可输入。其默认值是true。
- **autoFocus**：如果为true，将自动聚焦。
- **clearButtonMode**：枚举类型，可选值有'never'、'while-editing'、'unless-editing'、'always'。用于显示清除按钮。
- **maxLength**：能够输入的最长字符数。
- **enablesReturnKeyAutomatically**：如果值为true，表示没有文本时键盘是不能有返回键的。其默认值为false。
- **returnKeyType**：枚举类型，可选值有'default'、'go'、'google'、'join'、'next'、'route'、'search'、'send'、'yahoo'、'done'、'emergency-call'。表示软键盘返回键显示的字符串。
- **secureTextEntry**：如果为true，则像密码框一样隐藏输入内容。默认值为false。
- **onChangeText**：当文本输入框的内容变化时，调用该函数。onChangeText接收一个文本的参数对象。
- **onChange**：当文本变化时，调用该函数。
- **onEndEditing**：当结束编辑时，调用该函数。
- **onBlur**：失去焦点触发事件。
- **onFocus**：获得焦点触发事件。
- **onSubmitEditing**：当结束编辑后，点击键盘的提交按钮触发该事件。

3.4.2 案例：搜索自动提示

在不少应用中，都有基于位置的搜索，这里我们以高德地图App的搜索为例进行介绍，我们需要完成的是搜索框和自动提示的结果列表，如图3-13所示。

图3-13 地名搜索

在App中，搜索功能往往是一个组件，因此，将搜索功能封装成组件是一个很好的决定。我们按照以下步骤完成搜索功能。

1. 输入框

React Native默认的输入框还是很不美观的，因此，对输入框的定制是必不可少的。我们设计输入框距离屏幕左侧5pt，搜索按钮距离屏幕右侧5pt。这样，我们的搜索输入框左右距离是一致的，看起来比较协调。下面定义Search组件，具体代码如下：

```
var React = require('react-native');
var {
  StyleSheet,
  Text,
  AppRegistry,
  View,
  TextInput,
} = React;

var Search = React.createClass({
  render: function(){
    return (
      <View style={[styles.flex, styles.flexDirection]}>
        <View style={styles.flex}>
          <TextInput style={styles.input} returnKeyType="search"/>
        </View>
        <View style={styles.btn}>
          <Text style={styles.search}>搜索</Text>
        </View>
      </View>
    );
  }
});

var App = React.createClass({
  render: function(){
    return(
      <View style={[styles.flex, styles.topStatus]}>
        <Search></Search>
      </View>
    );
  }
});

var styles = StyleSheet.create({
  flex:{
    flex: 1,
  },
  flexDirection:{
    flexDirection:'row'
  },
  topStatus:{
    marginTop:25,
  },
```

```
    input:{
      height:45,
      borderWidth:1,
      marginLeft: 5,
      paddingLeft:5,
      borderColor: '#ccc',
      borderRadius: 4
    },
    btn:{
      width:55,
      marginLeft:-5,
      marginRight:5,
      backgroundColor:'#23BEFF',
      height:45,
      justifyContent:'center',
      alignItems: 'center'
    },
    search:{
      color:'#fff',
      fontSize:15,
      fontWeight:'bold'
    }

});

AppRegistry.registerComponent('APP', () => App);
```

运行程序，得到的输入框的效果如图3-14所示。

图3-14　搜索框

如果点击发现无法弹出虚拟键盘，可以到iOS Simulator（模拟器）上进行设置，具体可以参照图3-15。首先，选中Connect Hardware Keyboard，然后点击Toggle Software Keyboard即可。

图3-15　键盘设置

2. 自动提示列表

很多App的搜索都是这样的：当我们输入一个关键字的时候，会列出相关的搜索结果列表。一般情况下，完成该功能需要一个搜索服务，返回n条结果，然后将其展示出来。这里，我们使用静态数据模拟结果。我们的搜索结果需要根据用户是否输入关键字而显示。这里我们使用show变量标识是否显示结果列表。具体代码如下：

```
var React = require('react-native');
var {
  StyleSheet,
  Text,
  AppRegistry,
  View,
  TextInput,
} = React;
var onePT = 1 / React.PixelRatio.get();
var Search = React.createClass({
  getInitialState: function(){
    return {
      show: false
    };
  },
  getValue: function(text){
    var value = text;
    this.setState({
      show: true,
      value: value
    });
  },
  hide: function(val){
    this.setState({
      show: false,
      value: val
    });
```

```
    },
    render: function(){
      return (
        <View style={styles.flex}>
          <View style={[styles.flexDirection, styles.inputHeight]}>
            <View style={styles.flex}>
              <TextInput
                style={styles.input}
                returnKeyType="search"
                placeholder="请输入关键字"
                onEndEditing={this.hide.bind(this, this.state.value)}
                value={this.state.value}
                onChangeText={this.getValue}/>
            </View>
            <View style={styles.btn}>
              <Text style={styles.search} onPress={this.hide.bind(this,
                this.state.value)}>搜索</Text>
            </View>
          </View>
          {this.state.show?
            <View style={[styles.result]}>
              <Text onPress={this.hide.bind(this, this.state.value + '庄')}
                style={styles.item} numberOfLines={1}>{this.state.value}庄</Text>
              <Text onPress={this.hide.bind(this, this.state.value + '园街')}
                style={styles.item} numberOfLines={1}>{this.state.value}园街</Text>
              <Text onPress={this.hide.bind(this, 80 + this.state.value + '综合商店')}
                style={styles.item} numberOfLines={1}>80{this.state.value}综合商店</Text>
              <Text onPress={this.hide.bind(this, this.state.value + '桃')}
                style={styles.item} numberOfLines={1}>{this.state.value}桃</Text>
              <Text onPress={this.hide.bind(this, '杨林' + this.state.value + '园')}
                style={styles.item} numberOfLines={1}>杨林{this.state.value}</Text>
            </View>
            : null
          }
        </View>
      );
    },

});

var App = React.createClass({
  render: function(){
    return(
      <View style={[styles.flex, styles.topStatus]}>
        <Search></Search>
      </View>
    );
  }
});

var styles = StyleSheet.create({
  flex:{
    flex: 1,
  },
  flexDirection:{
```

```
    flexDirection:'row'
  },
  topStatus:{
    marginTop:25,
  },
  inputHeight:{
    height:45,
  },
  input:{
    height:45,
    borderWidth:1,
    marginLeft: 5,
    paddingLeft:5,
    borderColor: '#ccc',
    borderRadius: 4
  },
  btn:{
    width:55,
    marginLeft:-5,
    marginRight:5,
    backgroundColor:'#23BEFF',
    height:45,
    justifyContent:'center',
    alignItems: 'center'
  },
  search:{
    color:'#fff',
    fontSize:15,
    fontWeight:'bold'
  },
  result:{
    marginTop:onePT,
    marginLeft:5,
    marginRight:5,
    height:200,
    borderColor:'#ccc',
    borderTopWidth:onePT,
  },
  item:{
    fontSize:16,
    padding:5,
    paddingTop:10,
    paddingBottom:10,
    borderWidth:onePT,
    borderColor:'#ddd',
    borderTopWidth:0,
  }
});
AppRegistry.registerComponent('APP', () => App);
```

我们通过判断this.state.show来确定是否显示结果列表。如果this.state.show是true，则显示。如果是false，则隐藏（null对象不显示）。结果列表的规则是：输入关键字+预设关键字。这样的话，我们省去了服务端。同时，点击结果列表中的某一项，应该隐藏列表并且将结果显示在输入框中。onPress={this.hide.bind(this, this.state.value + '庄')}就是当用户点击时，将字符串结果拼接传入到hide函数中。

hide函数很简单，就是将this.state.show设置为false，这样会将结果列表隐藏起来了。因为状态的改变引起了视图的重新渲染，遇到this.state.show为false，就不渲染结果列表。同时，设置value为我们拼接好的结果字符串。我们需要对TextInput做一些处理，才能更好地符合预期。我们期待，当用户点击结果列表中的某一项时，结果会出现在搜索框中。我们在TextInput组件上增加了如下几个属性。

- **returnKeyType**：因为这里的应用场景是搜索，所以们虚拟键盘的返回键是search。
- **placeholder**：显示在输入前的占位符"请输入关键字"。
- **onEndEditing**：用户结束编辑时触发该事件，会将this.state.value值写入。这样，就能在搜索框中显示该值。
- **value**：通过this.state.value修改TextInput的value值。
- **onChangeText**：监听输入框值的变化，onChangeText获取的值作为字符串传入。

在初始化的时候，设置结果列表为隐藏，即在getInitialState中设置show为false。

3. 完成的效果

我们可以继续完善部分样式，使得功能更加符合产品的预期。比如，使用React.PixelRatio.get()来获取最小线宽。该实例的完整代码可以参考https://github.com/vczero/React-Native-Code中的第3章3.4节。最终完成的效果如图3-16所示。

图3-16 搜索自动提示

3.5 Touchable 类组件

React Native没有像Web开发那样可以给元素（组件）绑定click事件。在3.2节中，我们知道Text组件有onPress事件，可以给Text组件绑定触摸点击事件。为了像Text组件那样使得其他组件可以被点击，React Native提供了3个组件来做这件事。这3个组件称为"Touchable类组件"，具体如下所示。

- TouchableHighlight：高亮触摸。用户点击时，会产生高亮效果。
- TouchableOpacity：透明触摸。用户点击时，点击的组件会出现透明过渡效果。
- TouchableWithoutFeedback：无反馈性触摸。用户点击时，点击的组件不会出现任何视觉变化。

3.5.1 TouchableHighlight 组件

在Native App中，我们希望点击的时候会有一些视觉上的变化。这样，视觉的变化会告知我们已经点击过了，从而避免重复点击。TouchableHighlight组件的属性如下所示。

- **activeOpacity**：触摸时透明度的设置。
- **onHideUnderlay**：隐藏背景阴影时触发该事件。
- **onShowUnderlay**：出现背景阴影时触发该事件。
- **underlayColor**：点击时背景阴影效果的背景颜色。

这里我们通过示例演示一下TouchableHighlight组件的用法，相关代码如下：

```
var React = require('react-native');

var {
  StyleSheet,
  Text,
  AppRegistry,
  View,
  TouchableHighlight,
} = React;

var App = React.createClass({
  show: function(text){
    alert(text);
  },
  render: function(){
    return(
      <View style={[styles.flex]}>
        <View>
          <TouchableHighlight
            onPress={this.show.bind(this, 'React Native入门与实战')}
            underlayColor="#E1F6FF"
          >
            <Text style={styles.item}>React Native入门与实战</Text>
          </TouchableHighlight>
```

```
            <TouchableHighlight
              onPress={this.show.bind(this, '图灵出版社')}
              underlayColor="#E1F6FF"
            >
              <Text style={styles.item}>图灵出版社</Text>
            </TouchableHighlight>
          </View>
        </View>
      );
    }
  });

var styles = StyleSheet.create({
  flex:{
    flex: 1,
    marginTop:25,
  },
  item:{
    fontSize:18,
    marginLeft:5,
    color:'#434343'
  }
});

AppRegistry.registerComponent('APP', () => App);
```

在上面的代码中，我们在Text组件外面包了一个TouchableHighlight组件。同时，我们在TouchableHighlight组件上使用onPress监听用户点击事件。当用户点击时，会传入被点击的字符串。同时我们设置了点击的背景颜色是#E1F6FF。按cmd+R快捷键运行代码，得到的效果如3-17所示。

图3-17　TouchableHighlight运行效果

3.5.2 TouchableOpacity 组件

TouchableOpacity组件不用设置背景颜色，这样更加方便使用。该组件只有一个属性activeOpacity。这里我们通过一个简单的例子来看一下。将TouchableHighlight替换成TouchableOpacity，修改后的代码如下：

```
var App = React.createClass({
  show: function(text){
    alert(text);
  },
  render: function(){
    return(
      <View style={[styles.flex]}>
        <View>
          <TouchableOpacity
            onPress={this.show.bind(this, 'React Native入门与实战')}>
            <Text style={styles.item}>React Native入门与实战</Text>
          </TouchableOpacity>

          <TouchableOpacity
            onPress={this.show.bind(this, '图灵出版社')}>
            <Text style={styles.item}>图灵出版社</Text>
          </TouchableOpacity>

          <TouchableOpacity>
            <View style={styles.btn}>
              <Text style={{fontSize:25,color:'#fff'}}>按钮</Text>
            </View>
          </TouchableOpacity>
        </View>
      </View>
    );
  }
});
```

同时，我们增加了一个样式btn，该样式的内容如下：

```
btn:{
  marginLeft:30,
  marginTop:30,
  width:100,
  height:100,
  backgroundColor:'#18B4FF',
  justifyContent: 'center',
  alignItems: 'center',
  borderRadius: 50,
}
```

3.5.3 TouchableWithoutFeedback 组件

就像官网上所说的那样，除非有很充足的理由，你才会使用TouchableWithoutFeedback组件，一般不建议使用该组件。如果没有触摸反馈的话，就会像Web交互一样，而不是Native交互。

TouchableWithoutFeedback组件支持3个事件，具体如下所示。

- **onLongPress**：长按事件。
- **onPressIn**：触摸进入事件。
- **onPressOut**：触摸释放事件。

这里我们不过多介绍TouchableWithoutFeedback组件，有兴趣的读者可以参考https://facebook.github.io/react-native/docs/touchablewithoutfeedback.html#content。

3.6 Image 组件

一款App中既需要文本，又需要图片。就像HTML提供了img元素一样，React Native提供了Image组件。React Native的Image组件调用的图片的途径比较多，例如网络图片、本地磁盘图片、照相机的图片等。

3.6.1 Image 组件介绍

Image组件是我们常用的组件。学完其他组件再学习Image组件，感觉比较轻松。Image组件目前支持的属性如下所示。

- **resizeMode**：枚举类型，其值为cover、contain、stretch。表示图片适应的模式。
- **source**：图片的引用地址，其值为{uri: string}。如果是一个本地的静态资源，那么需要使用require('image!name')包裹。
- **defaultSource**：iOS支持的属性，表示默认的图片地址。如果网络图片加载完成，将取代defaultSource。
- **onLoad**：iOS支持的属性，加载成功时触发该事件。
- **onLoadEnd**：iOS支持的属性，不管是加载成功还是失败，都会触发该事件。
- **onLoadStart**：iOS支持的属性，加载开始时触发该事件。
- **onProgress**：iOS支持的属性，加载过程的进度事件。

3.6.2 加载网络图片

一般情况下，我们都使用网络图片，因为网络图片容易更新和修改。加载网络图片的方式是source={uri: 'http://xxx.com/xx.png'}。这里我们写一个简单的图片浏览器，该图片浏览器有两个按钮："上一张"和"下一张"。我们的设计稿如图3-18所示。

图3-18　图片浏览器设计稿

按照设计稿的需求，我们将该功能封装成功能组件。下面我们来完成该功能。

React Native不像Web开发那样会默认显示图片。在浏览器中，图片的大小一开始是0×0，当图片下载完成后，会按照图片的大小渲染，这时你就会看到图片的加载闪烁，但这是不好的用户体验。因此，在React Native中，Image组件的默认大小是0，是不显示图片的。我们需要给定图片的宽高或者知道图片的宽高比才能展示图片。因此，我们一开始就可以使用占位符来替代，这样就没有了Web的闪烁。这里，我们设置主体图片的模式为resizeMode="contain"，这样的话，我们的图片就会在指定大小内自适应缩放。source的值为{uri: this.state.imgs[this.state.count]}，其中uri是根据this.state.count变化的，this.state.count代表数组中的索引。具体的代码如下所示：

```
var React = require('react-native');
var {
  StyleSheet,
  Text,
  AppRegistry,
  View,
  Image,
  TouchableOpacity,
} = React;

var imgs = [
  'http://www.ituring.com.cn/bookcover/1442.796.jpg',
  'http://www.ituring.com.cn/bookcover/1668.553.jpg',
  'http://www.ituring.com.cn/bookcover/1521.260.jpg'
];
var MyImage = React.createClass({
```

```js
getInitialState: function(){
  var imgs = this.props.imgs;
  return {
    imgs: imgs,
    count: 0
  };
},
goNext: function(){
  var count = this.state.count;
  count ++;
  if(count < imgs.length){
    this.setState({
      count: count
    });
  }
},
goPreview: function(){
  var count = this.state.count;
  count --;
  if(count >= 0){
    this.setState({
      count: count
    });
  }
},
render: function(){
  return(
    <View style={[styles.flex]}>
      <View style={styles.image}>
        <Image style={styles.img}
               source={{uri: this.state.imgs[this.state.count]}}
               resizeMode="contain"
        />
      </View>
      <View style={styles.btns}>
        <TouchableOpacity onPress={this.goPreview}>
          <View style={styles.btn}>
            <Text>上一张</Text>
          </View>
        </TouchableOpacity>
        <TouchableOpacity onPress={this.goNext}>
          <View style={styles.btn}>
            <Text>下一张</Text>
          </View>
        </TouchableOpacity>
      </View>
    </View>
  );
}
});

var App = React.createClass({
  render: function(){
    return(
```

```
        <View style={[styles.flex, {marginTop:40}]}>
          <MyImage imgs={imgs}></MyImage>
        </View>
      );
    }
  });
  var styles = StyleSheet.create({
    flex:{
      flex: 1,
      alignItems:'center'
    },
    image:{
      borderWidth:1,
      width:300,
      height:200,
      borderRadius:5,
      borderColor:'#ccc',
      justifyContent:'center',
      alignItems:'center'
    },
    img:{
      height:150,
      width:200,
    },
    btns:{
      flexDirection: 'row',
      justifyContent: 'center',
      marginTop:20
    },
    btn:{
      width:60,
      height:30,
      borderColor: '#0089FF',
      borderWidth: 1,
      justifyContent: 'center',
      alignItems:'center',
      borderRadius:3,
      marginRight:20,
    },
  });
```

完成的效果如图3-19所示。

图3-19 图片浏览器

3.6.3 加载本地图片

在3.6.2节中，我们介绍了如何加载网络图片，这一节将学习如何使用本地图片。很多时候，我们想使用本地图片，但不希望搭建静态文件服务器。React Native提供了静态图片的调用方式：source={require('image!my-icon')}。

React Native提供本地图片的加载方式的同时，也给出了建议：

```
//好的加载方式
<Image source={require('image!my-icon')} />

//不好的加载方式
var icon = this.props.active ? 'my-icon-active' : 'my-icon-inactive';
<Image source={require('image!' + icon)} />

//好的加载方式
var icon = this.props.active ? require('image!my-icon-active') : require('image!my-icon-inactive');
<Image source={icon} />
```

React Native希望我们使用第一种和第三种方式来加载本地图片。因为在将来打包静态资源的过程中，第一种和第三种方案可以很容易找出静态资源，而不是运行时去分析静态资源。虽然第二种方案在当下也是可行的，但是React Native还是建议我们遵守第一种和第三种的代码规范。

要在项目中添加本地图片，Xcode提供了个简单的方式，那就是直接将图片拖入到Images.xcassets中，如图3-20所示。但是这需要重新启动模拟器才能起作用，才能看到效果。

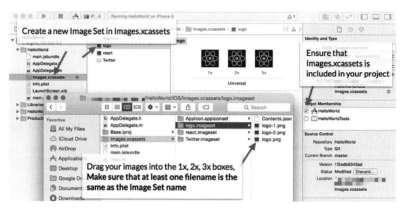

图3-20　加载本地图片

React Native计划后期让图片支持sprite(图片精灵或者雪碧图)，例如可以这样使用：{uri: ..., crop: {left: 10, top: 50, width: 20, height: 40}}。

3.7　TabBarIOS 组件

我们经常会用到的一个功能就是Tab切换。React Native就提供了该功能组件：TabBarIOS和TabBarIOS.Item。TabBarIOS组件就是为切换不同页面（视图或者路由）产生。因此，App中主体功能的切换基本上都是使用TabBarIOS组件。TabBarIOS.Item是TabBarIOS的附属组件。

3.7.1　TabBarIOS 组件介绍

TabBarIOS组件的属性比较少，主要有3个，如下所示。

- **barTintColor**：Tab栏的背景颜色。
- **tintColor**：当我们选中了某一个Tab时，该Tab的图标颜色。
- **translucent**：Tab栏是否透明。

TabBarIOS.Item组件是TabBarIOS组件的某一项Tab，支持如下属性。

- **badge**：红色的提示数字，可以用作消息提醒。
- **icon**：Tab的图标，如果不指定，默认显示系统图标。
- **onPress**：点击事件。当某个Tab被选中时，需要改变该组件的selected={true}设置。
- **selected**：是否选中某个Tab。如果其值为true，则选中并且显示子组件。
- **selectedIcon**：选中状态的图标，如果为空，则将图标变为蓝色。
- **systemIcon**：系统图标，其值是枚举类型，可选值有'bookmarks'、'contacts'、'downloads'、'favorites'、'featured'、'history'、'more'、'most-recent'、'most-viewed'、'recents'、'search'和'top-rated'。
- **title**：标题。它会出现在图标底部。当我们使用了系统图标时，将会忽略该标题。

3.7.2 案例：类 QQ Tab 切换

这里我们举一个常用的例子——QQ的Tab切换，如图3-21所示。

图3-21　QQ

这里我们需要完成QQ App的"消息"、"联系人"和"动态"的Tab切换。这里我们使用的图标可以到以下3个地址下载。

- "消息"图标：http://vczero.github.io/ctrip/message.png。
- "联系人"图标：http://vczero.github.io/ctrip/phone.png。
- "动态"图标：http://vczero.github.io/ctrip/star.png。

首先，加载TabBarIOS组件，然后使用TabBarIOS和TabBarIOS.Item进行布局。这里对于TabBarIOS.Item的图片，我们需要引用本地图片。React Native的TabBarIOS对图标进行了处理。当我们选择某个Tab时显示蓝色图标，而没有选择的图标会是灰色。这里我们不用管图片本来的颜色，TabBarIOS对图标的颜色进行了统一处理。这是一个很好的特性，用起来很方便。我们的具体代码如下：

```
var React = require('react-native');
var Dimensions = require('Dimensions');
var {
  StyleSheet,
  Text,
  AppRegistry,
  View,
  Image,
  ScrollView,
  TabBarIOS,
} = React;

var width = Dimensions.get('window').width;
var height = Dimensions.get('window').height - 70;
var App = React.createClass({
  getInitialState: function(){
    return {
      tab: 'message'
    };
  },
  select: function(tabName){
    this.setState({
      tab: tabName
    });
  },
  render: function(){
    return(
      <TabBarIOS style={styles.flex}>
        <TabBarIOS.Item
          title="消息"
          icon={require("image!message")}
          onPress={this.select.bind(this, 'message')}
          selected={this.state.tab === 'message'}>
          <ScrollView>
            <View style={styles.message}>
              <Text style={styles.message_title}>南山南</Text>
              <Text>
                他不再和谁谈论相逢的孤岛，因为心里早已荒芜人烟
                他的心里再装不下一个家，做一个只对自己说谎的哑巴，他说
```

```
                    你任何为人称道的美丽，不及他第一次遇见你
                    时光苟延残喘无可奈何
                    如果所有土地连在一起，走上一生只为去拥抱你
                    喝醉了他的梦，晚安
                    有天他听见有人唱着古老的歌，唱着今天还在远方发生的
                    像在她眼睛里看到的孤岛，没有悲伤但也没有花朵
                    你在南方的艳阳里大雪纷飞，我在北方的寒夜里四季如春
                    如果天黑之前来得及，我要忘了你的眼睛
                    穷极一生做不完一场梦
                    大梦初醒荒唐了这一生
                    南山南，北秋悲
                    南山有谷堆
                    南风喃 ，北海北
                    北海有墓碑
                </Text>
            </View>
        </ScrollView>
    </TabBarIOS.Item>
    <TabBarIOS.Item
        title="联系人"
        icon={require("image!phone")}
        onPress={this.select.bind(this, 'phonelist')}
        selected={this.state.tab === 'phonelist'}>
        <ScrollView>
            <Text style={styles.list}>
                <Text>唐三藏</Text>
                <Text>131-8904-9077</Text>
            </Text>
            <Text style={styles.list}>
                <Text>孙悟空</Text>
                <Text>131-8904-9078</Text>
            </Text>
            <Text style={styles.list}>
                <Text>猪八戒</Text>
                <Text>131-8904-9079</Text>
            </Text>
            <Text style={styles.list}>
                <Text>沙和尚</Text>
                <Text>131-8904-9080</Text>
            </Text>
        </ScrollView>
    </TabBarIOS.Item>
    <TabBarIOS.Item
        title="动态"
        icon={require("image!star")}
        onPress={this.select.bind(this, 'star')}
        selected={this.state.tab === 'star'}>
        <ScrollView style={styles.flex}>
            <Image style={{width:width, height:height}}
                source={{uri:'http://vczero.github.io/ctrip/star_page.jpg'}}/>
        </ScrollView>
    </TabBarIOS.Item>
</TabBarIOS>
);
```

```
      }
   });

   var styles = StyleSheet.create({
     flex:{
       flex: 1,
     },
     message:{
       alignItems:'center',
       marginLeft:5,
       marginRight:5,
     },
     message_title:{
       fontSize:18,
       color: '#18B5FF',
       marginBottom:5,
     },
     list:{
       height:30,
       fontSize:15,
       marginLeft:10,
       marginTop:10,
     }
   });
   AppRegistry.registerComponent('APP', () => App);
```

同时，我们在TabBarIOS.Item上添加了如下几个属性。

- **title**：表示Tab图标下面的小标题。
- **icon**：表示图标，这里使用require("image!message")引用。
- **selected**：表示当前Tab项是否选中，如果选中，图标将会变蓝色。

我们需要做Tab切换的效果就是改变TabBarIOS.Item中selected属性的值。如果selected为true，则选中当前的Tab。此外，我们在初始化的时候添加一个属性：this.state.tab。tab表示当前选中的TabBarIOS.Item的字符串名称。我们在selected上做一个简单的判断selected={this.state.tab === 'message'}，如果相等，则说明选中。一般我们通过点击来触发选中，因此需要添加onPress事件，传递不同的字符串来修改this.state.tab，从而触发选中。如果选中，TabBarIOS.Item的子节点将会显示出来，而其他的TabBarIOS.Item子节点将隐藏。同时我们使用Dimensions获取了屏幕的宽高，以便我们展示全屏图片。完整代码可以参考https://github.com/vczero/React-Native-Code项目的第3章3.7节。最终完成的简单效果如图3-22所示。

图3-22　TabBarIOS组件演示效果

3.8　WebView 组件

很多时候我们需要在App中嵌入一个活动页，那么什么样的方式最为合适呢？我们需要不定时上一个活动，又需要不定时结束一个活动。如果使用Native方式的话，需要更新App。这对于用户来说，是个很不好的体验。因此，WebView组件可以帮助我们解决这个问题。目前，主流的Hybird开发也是基于WebView来实现的，下面简要介绍这个组件。

3.8.1　WebView 组件介绍

目前，React Native的版本是v0.10.0-rc，WebView的功能基本能满足需求。目前，WebView组件支持的属性如下。

- **automaticallyAdjustContentInsets**：表示是否自动调整内部内容，其值为true或者false。
- **bounces**：回弹效果。如果其值为false，则内容拉到底部或者头部不回弹。其默认值为true。
- **contentInset**：内部内容偏移值，该值为一个JavaScript对象{top: number, left: number, bottom: number, right: number}。
- **html**：HTML代码字符串。如果传入了HTML代码字符串，则渲染该HTML代码。
- **injectedJavaScript**：注入的JavaScript代码，其值为字符串。如果加上了该属性，就会在WebView里面执行JavaScript代码。
- **onNavigationStateChange**：监听导航状态变化的函数。
- **renderError**：监听渲染页面出错的函数。

- **startInLoadingState**:是否开启页面加载的状态,其值为true或者false。
- **renderLoading**:WebView组件正在渲染页面时触发的函数,需要同startInLoadingState一起使用。当startInLoadingState为true时该函数才起作用。
- **scrollEnabled**:表示WebView里面页面是否能滚动,如果其值为true表示可以滚动,false表示禁止滚动。
- **onNavigationStateChange**:页面导航状态改变时,触发该事件监听。
- **scalesPageToFit**:按照页面比例和内容宽高比例自动缩放内容。

3.8.2 案例:使用 WebView 组件加载微博页面

我们可以使用WebView做一些简单的营销活动页面。但是这里我不好找一个营销页面,因为营销活动会不断变化,一下线演示就不起作用了。所以,这里使用的是新郎微博的Web App作为内嵌页面,主要为了展示WebView的用法。将新浪微博的Web App在手机上全屏显示,然后注入一句alert('欢迎使用React Native')的JavaScript代码,并且去掉上拉下拉回弹的效果,具体的代码如下所示:

```
var React = require('react-native');
var Dimensions = require('Dimensions');

var {
  AppRegistry,
  StyleSheet,
  Text,
  WebView,
  View,
} = React;

var width = Dimensions.get('window').width;
var height = Dimensions.get('window').height;

var App = React.createClass({
  render: function() {
    return (
      <View style={styles.container}>
        <WebView
          injectedJavaScript="alert('欢迎使用React Native')"
          bounces={false}
          url='http://weibo.com/vczero'
          style={{width:width,height:height}}>
        </WebView>
      </View>
    );
  }
});

var styles = StyleSheet.create({
  container: {
```

```
    flex: 1
  }
});

AppRegistry.registerComponent('APP', () => App);
```

这里我们使用Dimensions来获取设备的宽度和高度。我们将获取到的宽度和高度给WebView，这样就会全屏显示，最终的运行效果如图3-23所示。

图3-23　新浪微博

如果想加入自己的HTML代码片段，只需要使用html属性即可。例如我们，加载一个全屏的图片，可以使用

```
html='<div><img src="http://vczero.github.io/ctrip/star_page.jpg"/></div>'
```

得到一张大图，同时使用contentInset去掉留白，将scrollEnabled设为false来禁止滚动。具体的代码如下所示：

```
var React = require('react-native');
var Dimensions = require('Dimensions');

var {
  AppRegistry,
  StyleSheet,
  Text,
  WebView,
  View,
} = React;
```

```
var width = Dimensions.get('window').width;
var height = Dimensions.get('window').height;

var App = React.createClass({
  render: function() {
    return (
      <WebView
        contentInset={{left:-10,top:-28}}
        scrollEnabled={false}
        html='<div><img src="http://vczero.github.io/ctrip/star_page.jpg"/></div>'
        style={{width:width,height:height}}>
      </WebView>
    );
  }
});

AppRegistry.registerComponent('APP', () => App);
```

3.8.3 案例：新浪微博 OAuth 认证

很多时候我们需要调用开放平台，但是不想嵌入SDK，只需要数据。因此，选择WebView来获取accessToken是一个很好的选择，具体代码如下：

```
var React = require('react-native');
var {
  AppRegistry,
  StyleSheet,
  Text,
  View,
  WebView
} = React;

var appKey = '4263807830';
var callback = 'http://127.0.0.1:3000';
var url = 'https://api.weibo.com/oauth2/authorize?client_id=' +
  appKey + '&redirect_uri=' + callback;

module.exports = React.createClass({
  getInitialState: function(){
    return {
      code: null
    };
  },
  render: function(){
    return (
      <View style={styles.container}>
        {
          !this.state.code ?
            <WebView style={styles.container} url={url}
              onNavigationStateChange={this.navChange}/>
          :<Text>{this.state.code}</Text>
        }
```

```
        </View>
      );
    },
    navChange: function(state){
      var _that = this;
      if(state.url.indexOf(callback + "/?code=") > -1){
        var code = state.url.split('?code=')[1];
        //TODO: 这里可以使用code去交换accessToken
        _that.setState({
          code: code
        });
      }
    }
});

var styles = StyleSheet.create({
  container:{
    flex:1
  }
});
```

这里使用的是测试的key。关于OAuth接口的信息,可以前往新浪微博开放平台(http://open.weibo.com/)查看。我们在WebView上使用了onNavigationStateChange事件监听。onNavigationStateChange的作用是当发现浏览器地址改变时,触发事件。在navChange函数中,我们判断是否已经正确跳转到回调地址。如果跳转正确,则获取code,最后通过code去获取accessToken。新浪微博OAuth认证登录界面如图3-24所示。

图3-24　登录界面

第4章 常用API及其实践

第3章介绍了React Native的常用组件，运用这些组件可以完成很多我们需要的UI界面和功能。其实，在第3章的一些案例中，我们已经提前用过一些API。因此，只有配合使用React Native的常用组件和常用API，才能更好地开发应用程序。在这一章中，我们将着重介绍React Native的常用API。为了更容易理解和学习每个API，我们通过示例来介绍它们。

4.1 AppRegistry

每一个应用程序的运行都有一个入口文件或者入口函数，而在React Native中，AppRegistry就肩负着这样的责任。

4.1.1 AppRegistry 介绍

AppRegistry负责注册运行React Native应用程序的JavaScript入口。我们的应用程序的入口组件需要使用AppRegistry.registerComponent来注册。当注册完应用程序组件后，Native系统（Objective-C）就会加载jsbundle文件并且触发AppRegistry.runApplication运行应用。AppRegistry有以下的方法。

- registerConfig(config: Array<AppConfig>)：静态方法，注册配置。
- registerComponent(appKey: string, getComponentFunc: ComponentProvider)：注册入口组件。
- registerRunnable(appKey: string, func: Function)：注册函数监听。
- getAppKeys()：获取registerRunnable注册的监听键。
- runApplication(appKey: string, appParameters: any)：运行App。

4.1.2 AppRegistry 示例

在前3章中，我们都使用了AppRegistry.registerComponent函数。如：

```
AppRegistry.registerComponent('App', () => App);
```

其实，AppRegistry的用法就是这么简单。我们希望了解更多，因此，我们进一步探索。启动应用的时候，我们会在Xcode的日志输出栏中看到如下信息：

```
2015-09-04 16:22:24.884 [info][tid:com.facebook.React.JavaScript]
 'Running application "App" with appParams: {"rootTag":1,
 "initialProps":{}}. __DEV__ === true, development-level
 warning are ON, performance optimizations are OFF'
```

其实，上面的日志是由runApplication打印出来的。在代码中加入alert(AppRegistry.runApplication);，看一下runApplication函数的定义：

```
alert(AppRegistry.runApplication);
AppRegistry.registerComponent('App', () => App);
```

我们打印出来的函数定义如图4-1所示。

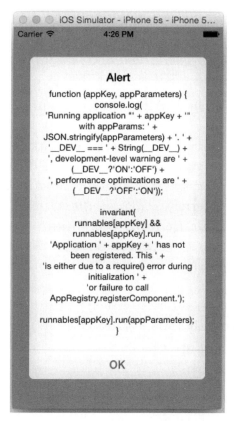

图4-1　runApplication函数的定义

可以看到，runApplication函数中console.log()打印的正是我们启动App时的日志。当然，也可以使用registerRunnable注册一些AppKey，示例代码如下：

```
AppRegistry.registerRunnable('vczero', function(){
  console.log('vczero');
});
AppRegistry.registerRunnable('react-native', function(){
  console.log('react-native');
});
alert(AppRegistry.getAppKeys());
```

4.2　AsyncStorage

AsyncStorage是一个简单的、具有异步特性的键值对的存储系统。相对整个App而言，它是全局的，应该用于替代LocalStorage。

4.2.1　AsyncStorage 介绍

AsyncStorage提供了比较全的方法供我们使用。每个方法都有一个回调函数，而回调函数的第一个参数都是错误对象。如果发生错误，该对象就会展示错误信息，否则为null。所有的方法执行后，都会返回一个Promise对象。具体的方法如下所示。

- `static getItem(key: string, callback:(error, result))`：根据键来获取值，获取的结果会在回调函数中。
- `static setItem(key: string, value: string, callback:(error))`：设置键值对。
- `static removeItem(key: string, callback:(error))`：根据键移除一项。
- `static mergeItem(key: string, value: string, callback:(error))`：合并现有值和输入值。
- `static clear(callback:(error))`：清除所有的项目。
- `static getAllKeys(callback:(error))`：获取所有的键。
- `static multiGet(keys, callback: (errors, result))`：获取多项，其中keys是字符串数组。
- `static multiSet(keyValuePairs, callback:(errors))`：设置多项，其中keyValuePairs是字符串的二维数组。
- `static multiRemove(keys, callback:(errors))`：删除多项，其中keys是字符串数组。
- `static multiMerge(keyValuePairs, callback:(errors))`：多个键值对合并，其中keyValuePairs是字符串的二维数组。

4.2.2　案例：购物车

在网上购物的时候，经常会用到购物车，我们需要将比较中意的商品添加到购物车，等我们需要付款的时候点击购物车即可购买。购物车不仅方便了客户存储需要购买的物品，而且在某种程度上给公司带来了流量和交易的可能。这里我们以购物车为例来说明AsyncStorage的用法。我们创建的原型图如图4-2所示。

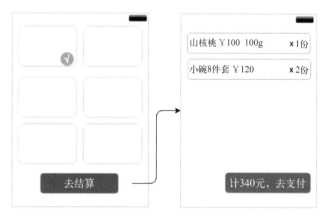

图4-2　购物车原型图

我们按照以下步骤来完成购物车。

1. 数据模型构建

这里以购买水果为例定义一个商品数组。其实，在真实的开发中，都是从服务端获取数据列表，这里我们使用静态数据，相关代码如下所示：

```
var Model = [
  {
    id: '1',
    title: '佳沛新西兰进口猕猴桃',
    desc: '12个装',
    price: 99,
    url: 'http://vczero.github.io/ctrip/guo_1.jpg'
  },
  {
    id:'2',
    title: '墨西哥进口牛油果',
    desc: '6个装',
    price: 59,
    url: 'http://vczero.github.io/ctrip/guo_2.jpg'
  },
  {
    id:'3',
    title: '美国加州进口车厘子',
    desc: '1000g',
    price: 91.5,
    url: 'http://vczero.github.io/ctrip/guo_3.jpg'
  },
  {
    id:'4',
    title: '新疆特产西梅',
    desc: '1000g',
```

```
      price: 69,
      url: 'http://vczero.github.io/ctrip/guo_4.jpg'
    },
    {
      id:'5',
      title: '陕西大荔冬枣',
      desc: '2000g',
      price: 59.9,
      url: 'http://vczero.github.io/ctrip/guo_5.jpg'
    },
    {
      id:'6',
      title: '南非红心西柚',
      desc: '2500g',
      price: 29.9,
      url: 'http://vczero.github.io/ctrip/guo_6.jpg'
    }
];
```

2. 列表项组件

列表项组件主要用于渲染商品的图片和名称，相关代码如下所示：

```
var Item = React.createClass({
  render: function(){
    return(
      <View style={styles.item}>
        <TouchableOpacity onPress={this.props.press}>
          <Image
            resizeMode="contain"
            style={styles.img}
            source={{uri:this.props.url}}>
            <Text numberOfLines={1} style={styles.item_text}>
              {this.props.title}</Text>
          </Image>
        </TouchableOpacity>
      </View>
    );
  }
});
```

这里我们在TouchableOpacity组件上绑定了点击事件，该事件由父组件传递。当点击时，该条商品就被添加进购物车中。Image组件上的resizeMode属性表示图片缩放的模型，这里我们使用"contain"，即图片会自适应在所在容器中。source属性表示图片的地址。同时，我们使用Text组件来展示商品的标题，该标题由父组件传递。这里使用了numberOfLines={1}，如果标题超过一行，则会以省略号（...）的形式展示剩余的文本。

3. 列表组件

列表组件会使用Item组件来渲染列表，并且列表是将两张图片并排显示，具体代码如下：

```
var List = React.createClass({
```

```
getInitialState: function(){
  return{
    count:0
  };
},
componentDidMount: function(){
  var _that = this;
  AsyncStorage.getAllKeys(function(err, keys){
    if(err){
      //TODO：存储取数据出错
      //给用户提示错误信息
    }
    //将存储的商品条数反应到按钮上
    _that.setState({
      count: keys.length
    });
  });
},
render: function() {
  var list = [];
  for(var i in Model){
    if(i % 2 === 0){
      var row = (
        <View style={styles.row} key={i}>
          <Item url={Model[i].url}
            title={Model[i].title}
            press={this.press.bind(this, Model[i])}></Item>
          <Item
            url={Model[parseInt(i)+1].url}
            title={Model[parseInt(i)+1].title}
            press={this.press.bind(this, Model[parseInt(i)+1])}></Item>
        </View>);
      list.push(row);
    }
  }

  var count = this.state.count;
  var str = null;
  if(count){
    str = '，共'+ count + '件商品';
  }
  return (
    <ScrollView style={{marginTop:10}}>
      {list}
      <Text onPress={this.goGouWu} style={styles.btn}>去结算{str}</Text>
    </ScrollView>
  );
},
goGouWu: function(){
  this.props.navigator.push({
    component: GouWu,
    title:'购物车'
  });
},
```

```
          press:function(data){
            var count = this.state.count;
            count ++;
            //改变数字状态
            this.setState({
              count: count
            });
            //AsyncStorage存储
            AsyncStorage.setItem('SP-' + this.genId() + '-SP', JSON.stringify(data), function(err){
              if(err){
                //TODO：存储出错
              }
            });
          },
          //生成随机ID：GUID
          //GUID生成的代码来自于Stoyan Stefanov
          genId:function(){
            return 'xxxxxxxx-xxxx-4xxx-yxxx-xxxxxxxxxxxx'.replace(/[xy]/g, function(c) {
              var r = Math.random() * 16 | 0,
              v = c == 'x' ? r : (r & 0x3 | 0x8);
              return v.toString(16);
            }).toUpperCase();
          }
        });
```

在上述代码中,我们在render方法中通过循环Model渲染列表项。这里使用i % 2 === 0来换行,并且给Item组件传递了url、title和press属性。在press方法中,我们将购物车商品的数量+1。同时,我们使用了AsyncStorage.setItem将选中的商品数据添加到App本地存储中。这里之所以使用SP-为前缀、-SP为后缀,采用GUID为存储的键名的一部分,是为了区分其他数据。这样做的好处有两个,具体如下所示。

❑ 可以区分用户数据,例如username等信息。
❑ 可以防止key值重复,保证同名商品都能被添加进购物车。

这样,我们就完成了商品信息的存储。

我们在componentDidMount方法中也做了一件比较重要的事,那就是当用户第二次进入App并且未支付的情况下,告诉用户目前购物车中有多少件商品。这里直接使用AsyncStorage.getAllKeys获取数据的条数。其实在实际开发中,需要根据前缀(SP-)和后缀(-SP)来判断是否是商品数据。

我们使用了Text组件作为"去结算"按钮。因为Text组件是可以绑定onPress事件的,所以我们绑定了goGouWu事件。这里使用this.props.navigator.push将购物车组件加载。

4. 购物车组件

完成了商品列表页的开发候,现在需要完成商品订单支付页,即购物车页面,具体代码如下:

```
var GouWu = React.createClass({
```

```
getInitialState: function(){
  return {
    data: [],
    price: 0
  };
},
render: function(){
  var data = this.state.data;
  var price = this.state.price;
  var list = [];
  for(var i in data){
    price += parseFloat(data[i].price);
    list.push(
      <View style={[styles.row, styles.list_item]}>
        <Text style={styles.list_item_desc}>
          {data[i].title}
          {data[i].desc}
        </Text>
        <Text style={styles.list_item_price}>¥{data[i].price}</Text>
      </View>
    );
  }
  var str = null;
  if(price){
    str = '，共' + price.toFixed(1) + '元';
  }
  return(
    <ScrollView style={{marginTop:10}}>
      {list}
      <Text style={styles.btn}>支付{str}</Text>
      <Text style={styles.clear} onPress={this.clearStorage}>清空购物车</Text>
    </ScrollView>
  );
},
componentDidMount: function(){
  var _that = this;
  AsyncStorage.getAllKeys(function(err, keys){
    if(err){
      //TODO:存储取数据出错
      //如果发生错误，这里直接返回（return）防止进入下面的逻辑
    }
    AsyncStorage.multiGet(keys, function(errs, result){
      //TODO：错误处理
      //得到的结果是二维数组
      //result[i][0]表示我们存储的键，result[i][1]表示我们存储的值
      var arr = [];
      for(var i in result){
        arr.push(JSON.parse(result[i][1]));
      }
      _that.setState({
        data: arr
      });
    });
```

```
      });
    },
    clearStorage: function(){
      var _that = this;
      AsyncStorage.clear(function(err){
        if(!err){
          _that.setState({
            data:[],
            price: 0
          });
          alert('购物车已经清空');
        }
        //TODO：ERR
      });
    }
  });
```

在购物车页面主要渲染用户选中的商品列表，该列表包含商品名称、价格等信息。因此，该组件也必须使用AsyncStorage API，才能获取我们在列表页存储的商品数据。同样，在componentDid-Mount方法中使用AsyncStorage.getAllKeys来获取AsyncStorage中的key。然后再使用AsyncStorage.multiGet获取商品数据的数组。这里result代表的是二维数组，result[i][0]表示存储的键，result[i][1]表示存储的值。我们将获取到的数据设置到this.state.data上。这样，当组件被加载完成后，就会显示商品列表。

此外，还需要完成清空购物车的功能。这里使用AsyncStorage.clear清除AsyncStorage中的所有数据。这里只是为了演示AsyncStorage的用法，所以没有完成支付功能，有兴趣的同学可以继续完善。

5. 完成整体功能

一切组件都已经开发完成了，现在需要做的就是将各组件串联起来，具体的使用方法如下：

```
var React = require('react-native');
var {
  AppRegistry,
  StyleSheet,
  Text,
  View,
  Image,
  NavigatorIOS,
  ScrollView,
  AsyncStorage,
  TouchableOpacity,
} = React;

//以下省略之前的代码，其中...表示省略的代码
var Model = [....];
var Item = React.createClass({...});
var List = React.createClass({...});
var GouWu = React.createClass({...});
```

```
var App = React.createClass({
  render: function() {
    return (
      <NavigatorIOS
        style={styles.container}
        initialRoute={
          {
            component: List,
            title: '水果列表'
          }
        }/>
    );
  }
});
var styles = StyleSheet.create({
  container: {
    flex: 1,
  },
  row:{
    flexDirection: 'row',
    marginBottom: 10,
  },
  item:{
    flex:1,
    marginLeft:5,
    borderWidth:1,
    borderColor:'#ddd',
    marginRight:5,
    height:100,
  },
  img:{
    flex:1,
    backgroundColor: 'transparent'
  },
  item_text:{
    backgroundColor: '#000',
    opacity: 0.7,
    color:'#fff',
    height:25,
    lineHeight:18,
    textAlign:'center',
    marginTop:74
  },
  btn:{
    backgroundColor:'#FF7200',
    height:33,
    textAlign:'center',
    color:'#fff',
    marginLeft:10,
    marginRight:10,
    lineHeight:24,
    marginTop:40,
```

```
      fontSize:18,
    },
    list_item:{
      marginLeft:5,
      marginRight:5,
      padding:5,
      borderWidth:1,
      height:30,
      borderRadius:3,
      borderColor:'#ddd'
    },
    list_item_desc:{
      flex:2,
      fontSize:15
    },
    list_item_price:{
      flex:1,
      textAlign:'right',
      fontSize:15
    },
    clear:{
      marginTop:10,
      backgroundColor:'#FFF',
      color:'#000',
      borderWidth:1,
      borderColor:'#ddd',
      marginLeft:10,
      marginRight:10,
      lineHeight:24,
      height:33,
      fontSize:18,
      textAlign:'center',
    }
});

AppRegistry.registerComponent('App', () => App);
```

这里，我们也使用了NavigatorIOS组件来做路由控制，初始化路由是List组件。整个购物车的完整代码可以参考https://github.com/vczero/React-Native-Code项目第4章4.2节。最后完成的效果如图4-3所示。

图4-3 购物车效果图

4.3 AlertIOS

在使用一款App的时候,经常会出现对话框,这在Web开发中通过alert实现。但是,alert比较丑陋,不符合我们的业务和审美要求,此时需要定制自己的对话框。React Native在这一点上做得很好,给我们提供了原生的对话框,那就是AlertIOS。

4.3.1 AlertIOS

AlertIOS组件应用很广,但是使用也十分简单,它的静态方法有如下两个。

- alert(title, message, buttons):普通对话框,其中buttons是对象数组。
- prompt(title, value, buttons):提供输入的对话框,其中buttons是对象数组。

butons的形式为:

```
[
    {
        text: '按钮显示的字符串',
        onPress: function(){
            //点击按钮触发的事件
        }
    }
]
```

默认的AlertIOS组件会提供一个"确认"（或者OK）按钮。默认情况下，数组中最后一个按钮高亮显示。如果数组的长度过长，按钮就会垂直排列。

4.3.2　AlertIOS 组件的应用

这里使用AlertIOS组件做一个简单的演示，其中会使用AlertIOS.alert函数简单地弹出对话框，使用AlertIOS.prompt弹出用户输入的值，具体的代码如下所示：

```
var React = require('react-native');
var {
  AppRegistry,
  StyleSheet,
  Text,
  View,
  AlertIOS,
} = React;

var App = React.createClass({
  render: function(){
    return(
      <View style={styles.container}>
        <Text style={styles.item} onPress={this.tip}>提示对话框</Text>
        <Text style={styles.item} onPress={this.input}>输入对话框</Text>
      </View>
    );
  },
  tip: function(){
    AlertIOS.alert('提示', '选择学习React Native',[
      {
        text: '取消',
        onPress: function(){
          alert('你点击了取消按钮');
        }
      },
      {
        text: '确认',
        onPress: function(){
          alert('你点击了确认按钮');
        }
      },
    ]);
  },
  input: function(){
    AlertIOS.prompt('提示', '使用React Native开发App',[
      {
        text: '取消',
        onPress: function(){
          alert('你点击了取消按钮');
        }
      },
      {
```

```
            text:'确认',
            onPress: function(e){
                alert(e);
            }
        },
    ]);
  }
});

var styles = StyleSheet.create({
  container:{
    flex:1,
    marginTop:25
  },
  item:{
    marginTop:10,
    marginLeft:5,
    marginRight:5,
    height:30,
    borderWidth:1,
    padding:6,
    borderColor:'#ddd'
  }
});

AppRegistry.registerComponent('App', () => App);
```

运行上述代码，得到的效果如图4-4所示。

图4-4　AlertIOS组件的演示效果

4.4 ActionSheetIOS

在App开发中我们也会遇到这样的需求，那就是分享和弹出多项选择操作。在iOS开发中，ActionSheet提供了这样的功能。而React Native同样封装了该功能，那就是ActionSheetIOS。

4.4.1 ActionSheetIOS 介绍

ActionSheetIOS提供了两个静态方法，具体如下所示。

- showActionSheetWithOptions(options, callback)：用于弹出分类菜单。
- showShareActionSheetWithOptions(options, failureCallback, successCallback)：分享弹出窗。

4.4.2 ActionSheetIOS 应用

ActionSheetIOS的用法同AlertIOS一样简单，这里我们简单演示一下，具体代码如下所示：

```
var React = require('react-native');
var {
  AppRegistry,
  StyleSheet,
  Text,
  View,
  ActionSheetIOS,
} = React;

var App = React.createClass({
  render: function(){
    return(
      <View style={styles.container}>
        <Text style={styles.item} onPress=
          {this.tip}>showActionSheetWithOptions</Text>
        <Text style={styles.item} onPress=
          {this.share}>showShareActionSheetWithOptions</Text>
      </View>
    );
  },
  tip: function(){
    ActionSheetIOS.showActionSheetWithOptions({
      options: [
        '拨打电话',
        '发送邮件',
        '发送短信',
        '取消'
      ],
      cancelButtonIndex: 3,
      destructiveButtonIndex: 0,
    },function(index){
```

```
        alert(index);
      });
    },
    share: function(){
      ActionSheetIOS.showShareActionSheetWithOptions({
        url: 'https://code.facebook.com',
      },function(err){
        alert(err);
      }, function(e){
        alert(e);
      });
    }
  });
  
  var styles = StyleSheet.create({
    container:{
      flex:1,
      marginTop:25
    },
    item:{
      marginTop:10,
      marginLeft:5,
      marginRight:5,
      height:30,
      borderWidth:1,
      padding:6,
      borderColor:'#ddd'
    }
  });

  AppRegistry.registerComponent('App', () => App);
```

从上面的代码可以看到showActionSheetWithOptions的使用方式。showActionSheetWith-Options的第一个参数是一个对象，该对象的options属性是字符串数组，表示可选项的名称。第二个参数cancelButtonIndex表示"取消"按钮的位置，即"取消"按钮在options数组中的索引。第三个参数destructiveButtonIndex表示不能使用的按钮位置，即不能使用的按钮在options数组中的索引。

showShareActionSheetWithOptions方法主要用于分享一个url，它的第一个参数是一个对象，第二个参数是失败的回调函数，第三个参数是成功的回调函数。

上述代码的运行效果如图4-5所示。

图4-5　ActionSheetIOS的演示效果

4.5　PixelRatio

随着技术的发展，很多设备都是高清屏，这在iOS移动设备上尤为明显。为了保证图片的显示效果一致，我们需要针对不同的设备准备多套图片，这是一件烦琐的事。实际上，在React Native中，pt单位是统一的，但是移动设备的像素密度是不一样的。React Native提供了PixelRatio来告知开发者像素密度。比如，iPhone 4的像素密度是2，那么在该设备上表达一个线宽就需要使用1/PixelRatio.get()。

4.5.1　PixelRatio 介绍

PixelRatio提供了一些静态方法供我们使用，具体如下。

❑ **get()**：获取像素密度。例如：

```
PixelRatio.get() === 1
  mdpi Android devices (160 dpi)
PixelRatio.get() === 1.5
  hdpi Android devices (240 dpi)
PixelRatio.get() === 2
  iPhone 4, 4S
  iPhone 5, 5c, 5s
  iPhone 6
```

```
  xhdpi Android devices (320 dpi)
PixelRatio.get() === 3
  iPhone 6 plus
  xxhdpi Android devices (480 dpi)
PixelRatio.get() === 3.5
  Nexus 6
```

- **getPixelSizeForLayoutSize(number)**：获取一个布局元素的像素大小，其返回值是一个四舍五入的整型。该方法的定义也比较简单，相关代码如下：

```
function getPixelSizeForLayoutSize(layoutSize){
  return Math.round(layoutSize * PixelRatio.get());
}
```

- **getFontScale()**：获取字体比例。在0.15.0版本中，目前只支持Android，iOS默认还是使用像素密度。该函数的定义如下：

```
function getFontScale(){
  return Dimensions.get('window').fontScale || PixelRatio.get();
}
```

4.5.2　PixelRatio 应用

这里我们使用两个简单的示例来说明PixelRatio的用途。

1. 最细边框

下面我们使用borderWidth:1/PixelRatio.get()来获取最细的边框：

```
var React = require('react-native');
var {
  AppRegistry,
  StyleSheet,
  Text,
  View,
  PixelRatio,
} = React;

var App = React.createClass({
  render: function(){
    return(
      <View style={styles.container}>
        <View style={{borderWidth:1, borderColor:'red', height:40,
          marginBottom:20}}></View>
        <View style={{borderWidth:1/PixelRatio.get(),
          borderColor:'red', height:40}}></View>
      </View>
    );
  },
});
```

```
var styles = StyleSheet.create({
  container:{
    flex:1,
    marginTop:25
  },
});

AppRegistry.registerComponent('App', () => App);
```

2. 自适应图片

我们希望一个图片既可以用在普通设备上，又能用在高清设备上，此时就需要获取图片的真实像素，这可以使用PixelRatio.getPixelSizeForLayoutSize来实现，实际上也就是Math.round(layoutSize * PixelRatio.get())。下面的代码只是示意代码，具体可以根据需求修改：

```
var image = getImage({
  width: PixelRatio.getPixelSizeForLayoutSize(300),
  height: PixelRatio.getPixelSizeForLayoutSize(200),
});
```

4.6 AppStateIOS

运行一款App的时候，需要知道该App的运行状态，这样我们可以在合适的时机（根据运行状态）做一些合理的事情。React Native提供了AppStateIOS来告知我们App的状态——激活状态（前台运行）和后台运行，甚至可以通知我们状态的改变。此外，AppStateIOS也经常用于做推送通知。

4.6.1 AppStateIOS 介绍

AppStateIOS拥有添加和删除事件的静态方法，因此，我们可以在代码中添加事件监听。以下是AppStateIOS的属性和事件。

- `currentState`：我们可以通过AppStateIOS.currentState获取当前App的属性。
- `addEventListener(type, handler)`：静态方法，用于添加事件监听。
- `removeEventListener(type, handler)`：静态方法，用于删除事件监听。

4.6.2 AppStateIOS 实例

我们既可以使用AppStateIOS.currentState获取应用的状态，也可以使用addEventListener来监听一些事件，比如change和memoryWarning事件。示例代码如下：

```
alert(AppStateIOS.currentState);

AppStateIOS.addEventListener('change', function(){
  //TODO：状态改变事件
```

```
});
AppStateIOS.addEventListener('memoryWarning', function(){
  //TODO：内存报警事件
});
```

4.7　StatusBarIOS

有时候，我们需要改变App的状态栏，这在React Native中通过StatusBarIOS来实现。

4.7.1　StatusBarIOS 介绍

StatusBarIOS有3个静态方法，具体如下所示。

- `setStyle(style, animated)`：设置状态栏的样式。其参数style是字符串，可以是'default'和'light-content'中的一个。animated是可选参数，表示是否有动画过渡，其值为true或者false。
- `setHidden(hidden, animated)`：用于隐藏状态栏。其参数hidden是boolean类型，如果其值为true，则隐藏状态栏，否则不隐藏。animated为可选参数，表示是否有动画过渡，其值为true或者false。
- `setNetworkActivityIndicatorVisible(visible)`：是否显示网络状态。其参数visible是boolean类型，如果其值为true，则显示网络状态，否则不显示。

4.7.2　StatusBarIOS 应用

首先，使用StatusBarIOS.setStyle来测试StatusBarIOS的用法。将整个容器的背景颜色设置为蓝色，这样我们可以看到状态栏变为白色，具体的代码如下：

```
var React = require('react-native');
var {
  AppRegistry,
  StyleSheet,
  Text,
  View,
  StatusBarIOS,
} = React;

StatusBarIOS.setStyle('light-content');
var App = React.createClass({
  render: function(){
    return(
      <View style={styles.container}></View>
    );
  },
});
```

```
var styles = StyleSheet.create({
  container:{
    flex:1,
    backgroundColor:'#1FB9FF'
  },
});

AppRegistry.registerComponent('App', () => App);
```

再使用setNetworkActivityIndicatorVisible来显示网络状态，具体的代码如下：

```
var React = require('react-native');
var {
  AppRegistry,
  StyleSheet,
  Text,
  View,
  StatusBarIOS,
} = React;

StatusBarIOS.setStyle('default');
StatusBarIOS.setNetworkActivityIndicatorVisible(true);
var App = React.createClass({
  render: function(){
    return(
      <View style={styles.container}></View>
    );
  },
});
var styles = StyleSheet.create({
  container:{
    flex:1
  },
});

AppRegistry.registerComponent('App', () => App);
```

整个代码的运行效果如图4-6所示。

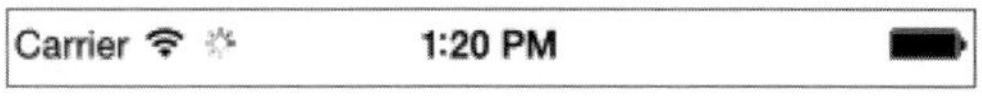

图4-6 状态栏

4.8 NetInfo

网络对一款App至关重要，我们需要根据网络的状态来采取不同的方案。比如，在WiFi网络的情况下，建议可以浏览大图；如果是离线状态，要关闭loading效果，及时提醒用户网络的状态，这样才能避免用户长时间等待。在React Native中，NetInfo API提供了获取网络状态的方法。

4.8.1　NetInfo 介绍

NetInfo 提供的属性和方法如下所示。

- `isConnected`：表示网络是否连接。
- `fetch()`：获取网络状态。
- `addEventListener(eventName, handler)`：添加事件监听。
- `removeEventListener(eventName, handler)`：删除事件监听。

其中，网络状态主要有以下几种类型。

- `none`：离线状态。
- `wifi`：在线状态，并且通过 WiFi 或者是 iOS 模拟器连接。
- `cell`：网络连接，通过 3G、WiMax 或者 LTE 进行连接。
- `unknown`：错误情况，网络状态未知。

4.8.2　NetInfo 示例

NetInfo 可以获取网络的状态和类型。比如，在手机 3G 网络的情况下建议用户开启"无图片浏览"；在 WiFi 的情况下，引导用户播放视频。获取网络状态的代码如下：

```
//获取连接类型
NetInfo.fetch().done(function(reachability){
  alert(reachability);
});

//获取是否连接
NetInfo.isConnected.fetch().done(function(isConnected){
  alert(isConnected);
});

//添加网络状态变化监听
NetInfo.addEventListener('change', function(reachability){
  alert(reachability);
});

//获取是否连接
NetInfo.isConnected.addEventListener('change', function(isConnected){
  alert(isConnected);
});
```

4.9　CameraRoll

现在 App 经常用的一个功能就是照相机，照相机的功能有很多：存储照片、获取照片、拍摄照片、拍摄视频、扫描二维码等。在 Web App 中，要使用这些功能，需要通过 Hybrid 实现。比如，在Web中获取照片的唯一ID就比较困难，但是在Native中却是件很容易的事。比如，我们有这样的一个需求——批量上传照片，上传照片的时候，需要将图片的base64和id传递到图片服务器。对于HTML5来说，我们需要定义自己的随机id，并且需要将随机id关联到本地资源，这是一件比较困难的事情。然后这在微信的JS-SDK中却能轻松完成。这是因为微信JS-SDK调用了Native的接口。我们使用React Native的一个好处就是：React Native提供了原生的API和组件，例如照相机的功能就是CameraRoll API。

4.9.1 CameraRoll 介绍

CameraRoll提供了对照相机控制的权限，它提供了两个比较简单的静态方法。

- `saveImageWithTag(tag, successCallback, errorCallback)`：用于保存图片到相册。参数tag是图片的地址，是字符串类型，在Android中是本地地址，例如"file:///sdcard/img.png"。在iOS中，tag的类型比较多，可以是url、assets-library、内存图片中的一种。successCallback是成功的回调函数，errorCallback是失败的回调函数。
- `getPhotos(params, callback, errorCallback)`：用于获取相册中的图片。params表示获取照片的参数，callback是成功回调，errorCallback是失败回调。

4.9.2 CameraRoll 应用

为了更好地说明CameraRoll的用法，这里用示例简单演示一下。同时，CameraRoll API的调用可以在模拟器上展示出来。这对于我们开发程序来说是一件幸福的事情，既不用打包成App，也不用将bundle文件上传到服务器。

该案例主要有两个功能——将网络图片保存到相册、显示相册中的图片，具体步骤如下：

1. 保存网络图片

很多时候，我们需要用到的一个功能就是将网络图片保存到本地，例如将微信朋友圈里好玩的图片存放到相册，这可以使用React Native的saveImageWithTag函数来实现。这里我们设置一个按钮，该按钮被点击时保存两张图片，相关代码如下：

```
//图片地址
var imgURL = 'http://vczero.github.io/lvtu/img/';
//视图
<View>
  <Text onPress={this.saveImg.bind(this, 'city.jpg', '3.jpeg')}
    style={[styles.saveImg]}>保存图片到相册</Text>
</View>
//保存功能
saveImg: function(img1, img2){
  var _that = this;
  CameraRoll.saveImageWithTag(imgURL + img1, function(url){
    if(url){
```

```
        var photos = _that.state.photos;
        photos.push(url);
        _that.setState({
          photos: photos
        });
        CameraRoll.saveImageWithTag(imgURL + img2, function(url){
          _that.setState({
            photos: photos
          });
          photos.push(url);
          alert('保存图片成功');
        }, function(){
          alert('保存图片失败');
        });
      }
    },function(){
      alert('保存图片失败');
    });
}
```

这里简单地保存两张图片，saveImageWithTag中的成功回调函数获取到的数据是被保存图片的ID。

2. 获取图片

有时候，我们需要显示相机里的图片。比如，发表微信或者上传图片时，需要在上传前预览图片，而不是直接将图片上传到服务器再返回图片数据进行预览。又比如，发表QQ状态时，会提示选择最近拍摄照片是否要上传。这个用户体验很好，因为我们希望拿到用户最近的拍摄图片。在React Native中，可以使用getPhotos来获取手机相册里的图片。这个示例的具体代码如下：

```
//参数
var fetchParams = {
  first: 5,
  groupTypes: 'All',
  assetType: 'Photos',
};
//获取照片
componentDidMount: function(){
  var _that = this;
  CameraRoll.getPhotos(fetchParams, function(data){
    var edges = data.edges;
    var photos = [];
    for(var i in edges){
      photos.push(edges[i].node.image.uri);
    }
    _that.setState({
      photos: photos
    });
  }, function(){
    alert('获取照片失败');
  });
},
```

在上述代码中，getPhotos的第一个参数是fetchParams，其定义如下：

```
var fetchParams = {
  first: 5, //获取数据的个数
  groupTypes: React.PropTypes.oneOf([ //数据的分组类型，可以是数组中的任意一个
    'Album',
    'All',
    'Event',
    'Faces',
    'Library',
    'PhotoStream',
    'SavedPhotos',
  ]),
  assetType: React.PropTypes.oneOf([ //资源类型，可以是数组中任意一个
    'Photos',
    'Videos',
    'All',
  ]),
};
```

其实，需要我们注意的是getPhotos成功回调时返回的数据，结果如下：

```
{
  edges: [
    {
      node: {
        timestamp: 1405312098,
        group_name: 'CameraRoll',
        type: 'ALAssetTypePhoto',
        image: {
          isStored: true,
          height: 669,
          uri: 'assets-library: //asset/asset.JPG?id=
            C9DB366F-350B-469E-88F7-76C785040206&ext=JPG',
          width: 1008
        },
        location: {}
      }
    },
    {
      node: {
        timestamp: 1405312098,
        group_name: 'CameraRoll',
        type: 'ALAssetTypePhoto',
        image: {
          isStored: true,
          height: 1001,
          uri: 'assets-library: //asset/asset.JPG?id=
            B6C0A21C-07C3-493D-8B44-3BA4C9981C25&ext=JPG',
          width: 1500
        },
        location: {}
      }
```

```
      },
      {
        node: {
          timestamp: 1405312097,
          group_name: 'CameraRoll',
          type: 'ALAssetTypePhoto',
          image: {
            isStored: true,
            height: 1250,
            uri: 'assets-library: //asset/asset.JPG?id=
              32D59830-62A0-445D-82B7-32C7C02F8897&ext=JPG',
            width: 834
          },
          location: {}
        }
      }
    ],
    page_info: {
      has_next_page: true,
      start_cursor: 'assets-library: //asset/asset.JPG?id=
        C9DB366F-350B-469E-88F7-76C785040206&ext=JPG',
      end_cursor: 'assets-library: //asset/asset.JPG?id=
        32D59830-62A0-445D-82B7-32C7C02F8897&ext=JPG'
    }
}
```

可以看到，我们设置需要返回3条数据，即edges返回的3个node。每个node是一个对象，其中包含图片的时间戳、组名、类型、图片的宽高、本地uri以及位置信息。同时，page_info告诉我们还存在下一页图片。因为这里设置一次取3张，所以就是3张一页。我们可以根据has_next_page来判断是否需要继续加载图片。

3. 实例验证

通过前面两步的描述，我们大体知道了如何使用CameraRoll。但是，我们更希望用一个完整的例子展示出来，这样更清楚如何将CameraRoll用于实际生产。该实例的综合代码如下：

```
var React = require('react-native');
var {
  AppRegistry,
  CameraRoll,
  Image,
  ListView,
  StyleSheet,
  View,
  Text,
  ScrollView,
} = React;

var fetchParams = {
  first: 4,
  groupTypes: 'All',
  assetType: 'Photos',
```

```
        };
        var imgURL = 'http://vczero.github.io/lvtu/img/';

        var App = React.createClass({
          getInitialState: function(){
            return{
              photos: null
            };
          },
          render: function(){
            var photos = this.state.photos || [];
            var photosView = [];
            for(var i = 0; i < 4; i += 2){
              photosView.push(
                <View style={styles.row}>
                  <View style={styles.flex_1}>
                    <Image
                      resizeMode="stretch"
                      style={[styles.imgHeight, styles.m5]}
                      source={{uri:photos[i]}}/>
                  </View>
                  <View style={styles.flex_1}>
                    <Image
                      resizeMode="stretch"
                      style={[styles.imgHeight, styles.m5]}
                      source={{uri:photos[parseInt(i) + 1]}}/>
                  </View>
                </View>
              );
            }

            return(
              <ScrollView>
                <View style={styles.row}>
                  <View style={styles.flex_1}>
                    <Image
                      resizeMode="stretch"
                      style={[styles.imgHeight, styles.m5]}
                      source={{uri: imgURL + 'city.jpg'}}/>
                  </View>
                  <View style={styles.flex_1}>
                    <Image
                      resizeMode="stretch"
                      style={[styles.imgHeight, styles.m5]}
                      source={{uri: imgURL + '3.jpeg'}}/>
                  </View>
                </View>
                <View>
                  <Text onPress={this.saveImg.bind(this, 'city.jpg', '3.jpeg')}
                        style={[styles.saveImg]}>保存图片到相册</Text>
                </View>
                <View style={[{marginTop:20}]}>
                  {photosView}
                </View>
```

```
      </ScrollView>
    );
  },
  componentDidMount: function(){
    var _that = this;
    CameraRoll.getPhotos(fetchParams, function(data){
      console.log(data);
      var edges = data.edges;
      var photos = [];
      for(var i in edges){
        photos.push(edges[i].node.image.uri);
      }
      _that.setState({
        photos: photos
      });
    }, function(){
      alert('获取照片失败');
    });
  },

  saveImg: function(img1, img2){
    var _that = this;
    CameraRoll.saveImageWithTag(imgURL + img1, function(url){
      if(url){
        var photos = _that.state.photos;
        photos.unshift(url);
        _that.setState({
          photos: photos
        });
        CameraRoll.saveImageWithTag(imgURL + img2, function(url){
          photos.unshift(url);
          _that.setState({
            photos: photos
          });
          alert('保存图片成功');
        }, function(){
          alert('保存图片失败');
        });
      }
    },function(){
      alert('保存图片失败');
    });
  }
});

var styles = StyleSheet.create({
  flex_1:{
    flex:1
  },
  m5:{
    marginLeft:5,
    marginRight:5,
    borderWidth:1,
    borderColor:'#ddd'
```

```
    },
    row:{
      flexDirection: 'row'
    },
    imgHeight:{
      height:120
    },
    saveImg:{
      flex:1,
      height:30,
      textAlign:'center',
      marginTop:20,
      backgroundColor:'#3BC1FF',
      color:'#FFF',
      lineHeight:20,
      borderRadius:5,
      marginLeft:5,
      marginRight:5,
      fontWeight:'bold'
    }
});

AppRegistry.registerComponent('App', () => App);
```

运行上述代码,得到的效果如图4-7所示,点击"保存图片到相册"按钮,将该按钮上方的两张图片保存到本地,同时显示最近的4张图片。

图4-7　CameraRoll示例

接着在模拟器上使用快捷键shift+cmd+H回退到主页。然后点击Photos（即相册），查看保存的图片，如图4-8所示。

图4-8　在Photos中查看保存的图片

4.9.3　react-native-camera

很多时候，最基础的API往往不能满足功能需求，需要我们自己拓展和开发。但是因为项目紧张，我们又不想造轮子，因此需要寻找第三方库。如果第三方库能满足我们的全部需求最好，如果不能，可以修改和拓展第三方库。

这里我们需要调用摄像头。我们在GitHub上找到了一个开源库react-native-camera，地址为https://github.com/lwansbrough/react-native-camera。下面按照如下步骤来使用react-native-camera。

1. 安装react-native-camera

进入项目的根目录，即package.json的目录。打开终端，使用npm安装最新版本，如果没有错误，则说明安装成功：

```
npm install react-native-camera@latest --save
```

2. 添加工程

因为该第三方库不是单纯的JavaScript库，其中包含了Objective-C的代码，所以需要编译。将react-native-camera添加到我们的项目中，具体步骤如下所示。

(1) 打开Xcode，找到项目目录，右击Libraries，选择Add Files to [your project's name]。

(2) 选择node_modules下我们刚才安装的react-native-camera目录，进入该项目，选择并添加RCTCamera.xcodeproj项目文件，这时候你应该看到Libraries包含了react-native-camera项目。

(3) 选中该项目，在Xcode中间的设置栏中找到Build Phases → Link Binary With Libraries。点击"添加"按钮，选择libRCTCamera.a并添加。

(4) 在Libraries中找到并点击RCTCamera.xcodeproj。选择Build Settings选项卡，此时默认显示的是Basic，这里需要选中All选项卡。然后找到Header Search Paths项并点击，确保该项的值包含$(SRCROOT)/../react-native/React和$(SRCROOT)/../../React这两项，并且它们的值是recursive，如图4-9所示。

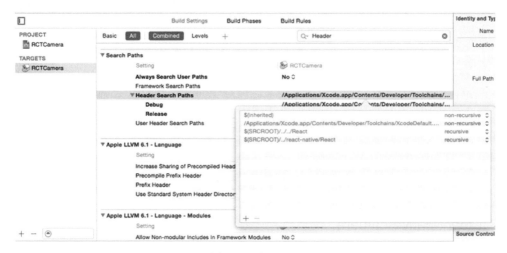

图4-9　项目配置

到此，我们完成了第三方库的添加。

3．使用react-native-camera

这时候使用cmd+R重启项目，会发现在模拟器中无法调用摄像头。我们需要将其打包后放在真机上运行才行，具体的打包流程可参见第7章。我们使用GitHub上的例子，该例子包含两个功能：前置摄像头和后置摄像头的切换、拍照。感谢lwansbrough提供该库，具体的代码如下：

```
var React = require('react-native');
var {
  AppRegistry,
  StyleSheet,
  Text,
  View,
  TouchableHighlight
} = React;
var Camera = require('react-native-camera');
```

```
var cameraApp = React.createClass({
  getInitialState() {
    return {
      cameraType: Camera.constants.Type.back
    }
  },

  render() {
    return (
      <Camera
        ref="cam"
        style={styles.container}
        onBarCodeRead={this._onBarCodeRead}
        type={this.state.cameraType}>
        <Text style={styles.welcome}>
          Welcome to React Native!
        </Text>
        <Text style={styles.instructions}>
          To get started, edit index.ios.js{'\n'}
          Press Cmd+R to reload
        </Text>
        <TouchableHighlight onPress={this._switchCamera}>
          <Text>The old switcheroo</Text>
        </TouchableHighlight>
        <TouchableHighlight onPress={this._takePicture}>
          <Text>Take Picture</Text>
        </TouchableHighlight>
      </Camera>
    );
  },
  _onBarCodeRead(e) {
    console.log(e);
  },
  _switchCamera() {
    var state = this.state;
    state.cameraType = state.cameraType === Camera.constants.Type.back
      ? Camera.constants.Type.front : Camera.constants.Type.back;
    this.setState(state);
  },
  _takePicture() {
    this.refs.cam.capture(function(err, data) {
      console.log(err, data);
    });
  }
});

var styles = StyleSheet.create({
  container: {
    flex: 1,
    justifyContent: 'center',
    alignItems: 'center',
    backgroundColor: 'transparent',
  },
```

```
  welcome: {
    fontSize: 20,
    textAlign: 'center',
    margin: 10,
  },
  instructions: {
    textAlign: 'center',
    color: '#333333',
  },
});

AppRegistry.registerComponent('App', () => cameraApp);
```

在真机上运行,可以看到效果。其实,基于react-native-camera可以拓展一个二维码的功能,这里就不详细介绍了。

4.10 VibrationIOS

使用微信的"摇一摇"功能时,会发出声音,这是对用户的一个听觉的提醒。如果关闭了声音,会通过"振动"表示我们摇到了一首歌曲或者一个电视频道。React Native在iOS上通过VibrationIOS API来实现这个功能。

VibrationIOS API只有一个方法,那就是vibrate()。如果调用vibrate()方法,移动设备就会出现1秒钟的振动效果。如果该设备不支持VibrationIOS,也不会出现负作用。

在模拟器中,我们是无法模拟振动效果的,需要在真机上才能体验到真实效果。具体的代码很简单,如下所示:

```
var React = require('react-native');
var {
  AppRegistry,
  StyleSheet,
  Text,
  View,
  VibrationIOS
} = React;

var App = React.createClass({
  render: function(){
    return(
      <View>
        <Text onPress={this.vibration} style={styles.btn}>振动一下</Text>
      </View>
    );
  },
  vibration: function(){
    VibrationIOS.vibrate();
  }
});
```

```
var styles = StyleSheet.create({
  btn:{
    marginTop:50,
    marginLeft:10,
    marginRight:10,
    height:35,
    backgroundColor:'#3BC1FF',
    color:'#fff',
    lineHeight:24,
    fontWeight:'bold',
    textAlign:'center'
  }
});
AppRegistry.registerComponent('App', () => App);
```

4.11　Geolocation

现在是移动互联网时代，移动设备越来越多，App也越来越丰富。基于位置的服务（LBS）也成为了App必不可少的功能。目前，国内高德地图、百度地图就提供了LBS服务，国外Google也提供了该项服务。LBS最基础的是获取位置信息，因此Geolocation（地理定位）显得十分重要。React Native也对Geolocation进行了封装，提供了Geolocation API。

4.11.1　Geolocation 介绍

Geolocation的功能如同HTML5 Geolocation一样，用过HTML5 Geolocation的同学对此应该十分熟悉。我们需要在项目的Info.list中添加NSLocationWhenInUseUsageDescription的key，才能开启该功能。当我们使用react-native init命令创建React Native的项目时，默认会开启地理定位功能。Geolocation的实现标准可以参考 https://developer.mozilla.org/en-US/docs/Web/API/Geolocation。

Geolocation提供的静态方法如下。

- **getCurrentPosition(successCallback, errorCallback, GeoOptions)**：用于获取当前的位置。其参数successCallback是成功回调函数，errorCallback是失败回调函数，GeoOptions是传递的参数。当前，GeoOptions支持的属性有：timeout（指定获取地理位置信息时的超时时长，其单位为毫秒。如果超时，则触发失败回调函数）、maximumAge（重复获取定时时指定多久再次获取，其单位为毫秒）、enableHighAccuracy（其值为true或者false，指定是否要求高精度的地理位置信息）。
- **watchPosition(successCallback, errorCallback, GeoOptions)**：监测位置运动。
- **clearWatch(watchID)**：依据ID清除监测。

4.11.2 Geolocation 应用

尽管我们可以在模拟器上调用Geolocation，但是在模拟器的定位十分不准。因此，建议打包后在真机上测试。我们使用var Geolocation = require('Geolocation')加载Geolocation API，具体的代码如下：

```
var React = require('react-native');
var {
  AppRegistry,
  StyleSheet,
  Text,
  View,
} = React;

var Geolocation = require('Geolocation');

var App = React.createClass({
  render: function(){
    return(
      <View>
        <Text onPress={this.vibration} style={styles.btn}>获取位置</Text>
      </View>
    );
  },
  vibration: function(){
    Geolocation.getCurrentPosition(function(data){
      alert(JSON.stringify(data));
    }, function(){
      alert('获取位置失败');
    });
  }
});

var styles = StyleSheet.create({
  btn:{
    marginTop:50,
    marginLeft:10,
    marginRight:10,
    height:35,
    backgroundColor:'#3BC1FF',
    color:'#fff',
    lineHeight:24,
    fontWeight:'bold',
    textAlign:'center'
  }
});

AppRegistry.registerComponent('catapp', () => App);
```

打包后在真机上运行的效果如图4-10所示。

图4-10 地理定位

目前，我是在上海跑的示例。根据上面的经纬度来看，定位相对还是很准的。我们可以看到，成功回调返回的结果如下所示：

```
{
  coords: {
    speed: -1,
    longitude: 121.3455563970114, //经度
    latitude: 31.21311039307059, //纬度
    altitude: 13.2847, //高程
    accuracy: 65,
    heading: -1,
    altitude: 0,
    altitudeAccuracy: 11.177
  },
  timestamp: 463765121483.258
}
```

其中longitude和latitude就是我们需要的经度和纬度。

4.12 数据请求

打开微信朋友圈，下拉刷新可以获取更多的朋友状态；打开微博，可以看到最新的微博；打开支付宝，向朋友发一个红包，这些都是互联网的作用和联系。在Web开发中，之所以能够请求和提交数据，是因为浏览器实现了一套网络API，以便我们可以通过浏览器和服务端交互。同样，

在App开发中,网络也是必不可少的。React Native同样需要实现网络API。但是,React Native遵循了浏览器的实现方式,实现了XMLHttpRequest API。

4.12.1 XMLHttpRequest

做Web开发的同学看到XMLHttpRequest都会感觉十分亲切,这是因为React Native对XMLHttpRequest的实现几乎和Web一样。它们的唯一区别就是:React Native中,XMLHttpRequest API不存在跨域的限制,而是作为全局API实现的。这里请求百度(http://www.baidu.com/)的首页,可以看到整个页面被抓取下来了,具体代码如下:

```
_doXMLHttpRequest: function(){
  var request = new XMLHttpRequest();
  request.onreadystatechange = (e) => {
    if (request.readyState !== 4) {
      return;
    }
    if (request.status === 200) {
      console.log('success', request.responseText);
    } else {
      console.warn('error');
    }
  };
  request.open('GET', 'http://www.baidu.com/');
  request.send();
}
```

当然,我们也可以封装成Zepto Ajax类型的API。同样,你也可以使用第三方的Ajax库来代替,例如superagent(https://github.com/visionmedia/superagent)。关于XMLHttpRequest的实现标准,可以参考https://developer.mozilla.org/en-US/docs/Web/API/XMLHttpRequest。

4.12.2 Fetch

相对XMLHttpRequest来说,Fetch是一个封装程度更高的网络API,它已经通过了标准委员会并在Chrome中实现。在React Native中,默认实现了Fetch。如果你想了解更多关于Fetch的内容,可以参考https://fetch.spec.whatwg.org/。 例如,我们可以将XMLHttpRequest获取百度首页数据的代码使用Fetch代替,具体代码如下所示:

```
_doFetch: function(){
  fetch('http://www.baidu.com/')
  .then(function(data){
    return data.text();
  })
  .then((responseText) => {
    console.log(responseText);
  })
  .catch((error) => {
    console.warn(error);
```

这里我们希望能够使用Fetch封装一个支持POST请求的简单API，具体代码如下所示：

```
function postRequest (url, data, callback) {
  var opts = {
    method: 'POST',
    headers: {
      'Accept': 'application/json',
      'Content-Type': 'application/json'
    },
    body: JSON.stringify(data)
  };

  fetch(url, opts)
  .then((response) => response.text())
  .then((responseText) => {
    callback(JSON.parse(responseText));
  });
}
```

本节的完整代码请参考https://github.com/vczero/React-Native-Code的第4章4.12节。

4.13 定时器

在这一节中，我们将介绍定时器API——setTimeout、setInterval、setImmediate和requestAnimationFrame，这些都是遵循浏览器API标准实现的，具体可以参考https://developer.mozilla.org/en-US/Add-ons/Code_snippets/Timers。在这一节中，我们只是概要介绍每个API，而不会具体介绍其用法，因为Web开发者对这些API都十分熟悉。

4.13.1 setTimeout

setTimeout主要用于设定一个定时任务，比如打开App 5秒后开始获取用户的位置信息，具体代码如下：

```
var timeoutID = setTimeout(function(){
  var geo = require('Geolocation');
  geo.getCurrentPosition(function(data){
    alert(JSON.stringify(data));
  }, function(e){
    alert(JSON.stringify(e));
  });
  if(timeoutID){
    clearTimeout(timeoutID);
  }
}, 5000);
```

4.13.2　setInterval

setInterval主要用于设定循环执行的任务，例如轮播图。这里，我们使用setInterval每隔5秒钟去请求百度的首页数据，具体代码如下：

```
setInterval(function(){
  fetch('http://www.baidu.com/')
    .then(function(data){
      return data.text();
    })
    .then((responseText) => {
      console.log(responseText);
    })
    .catch((error) => {
      console.warn(error);
    });
}, 5000);
```

4.13.3　setImmediate

setImmediate主要用于设置立即执行的任务，比如我们希望程序启动后立即发送日历到服务端，便于统计数据：

```
setImmediate(function(){
  console.log('发送数据');
});
```

4.13.4　使用 requestAnimationFrame 开发进度条

我们可以使用setInterval和setTimeout来实现动画，但是HTML5的出现，我们又多了两个选择，那就是CSS3动画和使用requestAnimationFrame实现动画。在React Native中，这可以通过requestAnimationFrame API来实现。

相对setTimeout(fn, 0)来说，requestAnimationFrame具有很强的优势：能够在动画流刷新后执行，即上一个动画流会完整执行。通常，我们会使用递归和requestAnimationFrame来实现动画。这里，我们使用React Native开发个简单的进度条，具体代码如下：

```
var React = require('react-native');
var {
  AppRegistry,
  StyleSheet,
  Text,
  View,
} = React;

var TimerMixin = require('react-timer-mixin');
var App = React.createClass({
  mixins: [TimerMixin],
```

```
    getInitialState: function(){
      return{
        width: 10
      };
    },
    render: function(){
      var css = [];
      css.push(styles.progress);
      if(this.state.width){
        css.push({width: this.state.width, marginTop:50});
      }
      return(
        <View>
          <View style={css}></View>
        </View>
      );
    },
    componentDidMount: function(){
      var _that = this;
      function doAnimated(){
        _that.setState({
          width: _that.state.width + 10
        });
        if(_that.state.width < 290){
          requestAnimationFrame(doAnimated);
        }
      }
      requestAnimationFrame(doAnimated);
    }
});
var styles = StyleSheet.create({
  progress:{
    height:10,
    width:10,
    marginLeft:10,
    backgroundColor:'#E72D00',
    marginTop:10
  }
});
AppRegistry.registerComponent('App', () => App);
```

这里，我们需要引入TimerMixin。我们需要通过npm来安装它：

```
npm i react-timer-mixin --save
```

4.13.5 完整代码

为了更好地展示以上定时器函数的用法，本书提供了一个完整的例子供大家参考，可前往 https://github.com/vczero/React-Native-Code项目第4章4.13节查看。

第 5 章
Native 扩展

移动终端的发展是爆炸式的,比如苹果移动设备,一年一个大版本的系统更新,持续地引入了 Touch ID、3D Touch 等全新的功能以及 API。同时在开发者社区中,开发者也在不断贡献代码,现在 GitHub 上已经托管了三万六千多个活跃的 Objective-C 项目。在这样的背景下,选择使用 React Native 开发移动应用时,难免遇到以下情况。

- 需要使用 React Native 未封装的原生功能。
- 重用已有的原生组件或者第三方组件。
- 多线程调用以及高性能要求的功能,例如加密、数据库、图形处理等。

对于这些问题,Facebook 自然不会毫无准备,React Native 提供了非常简单的规范,允许开发者自行根据需求来扩展原生组件。本章将介绍如何创建一个自定义的 Native 模块。

5.1 通信机制

React Native 作为一个 JavaScript 跨终端框架,与终端原生语言之间的交互是必不可少的。在 React Native 推出之前,如果想用 JavaScript 来开发,通常会选择混合模式 Hybrid,其最典型的框架就是 Cordova。它同样提供了 JavaScript 类库来调用原生的设备功能,这类应用使用 UIWebView 作为主要界面,通过向 WebView 请求加载特殊 URL 的方式来调用原生功能,最后调用 UIWebView 的 stringByEvaluatingJavaScriptFromString 接口执行 JavaScript 脚本传递数据。React Native 中的交互与上述 Hybrid 应用有着相似之处,React Native 使用 JavaScriptCore 框架提供的 JavaScript 运行环境,通过调用 evaluateScript 方法执行特定的 JavaScript 脚本来传递数据,完成 Objective-C 和 JavaScript 代码之间的交互。与 Hybrid 应用相比,React Native 中 JavaScript 构建的界面完全基于原生组件,原生界面的用户体验优于 UIWebView 则是毫无疑问的。当然,这也是利用 JavaScript 与原生语言的交互来实现的。

5.1.1 模块配置映射

在深入介绍细节之前,我们不妨思考一下,在编码中调用一个功能需要什么。很简单,这需要知道实现功能的是哪一个类的哪一个方法，需要传入什么参数即可。React Native也是如此，当JavaScript调用Native模块时，必须包含模块、方法和参数的信息，从而定位具体调用了哪个模块的哪个方法。此时需要一个模块的配置信息映射表，来作为双方通信的桥梁。现在，第一个问题就是React Native如何提供这个模块配置映射表。

在React Native加载的时候，所有注册并且符合规范的模块都会被导出，生成对应的模块数据类RCTModuleData，而模块数据中缓存了模块的对象实例，以及模块索引ModuleID：

```objc
//RCTBatchedBridge.m

- (void)initModules {
  ...
  module DataByID = [NSMutableArray new];
  for (Class moduleClass in RCTGetModuleClasses()) {
    NSString* moduleName = RCTBridgeModuleNameForClass(moduleClass);
    ...
    //实例化模块类
    module = [moduleClass new];
    ...
    modulesByName[moduleName] = module;
    ...
  }
  ...
  for (id<RCTBridgeModule> module in _modulesByName.allValues) {
    ...
    //生成索引ModuleID，从0开始计数
    //RCTModuleData对象缓存了ModuleID、Module实例和JavaScript执行器
    RCTModuleData *moduleData = [[RCTModuleData alloc]
      initWithExecutor:_javaScriptExecutor
      moduleID:@(_moduleDataByID.count)
      instance:module];
    [_moduleDataByID addObject:moduleData];
  }
  ...
}
```

在RCTModuleData模块数据中，当前模块所有的模块方法都会通过Objective-C运行时方法被导出，生成模块方法类RCTModuleMethod。而RCTModuleMethod对象缓存了原生方法与JavaScript方法的对应关系以及模块的实例，并且在生成配置的时候会顺序生成索引MethodID：

```objc
//RCTModuleData.m

- (NSArray *)methods
{
  ...
  NSArray *entries = ((NSArray *(*)(id,SEL))imp)(_moduleClass,selector);
  RCTModuleMethod *moduleMethod =
    [[RCTModuleMethod alloc] initWithObjCMethodName:entries[1]
                                       JSMethodName:entries[0]
                                        moduleClass:_moduleClass];
```

```
    [moduleMethods addObject:moduleMethod];
    ...
}

- (NSDictionary *)config
{
    ...
    methodconfig[method.JSMethodName] = @{
      @"methodID": @(idx),
      @"type": method.functionKind ==
        RCTJavaScriptFunctionKindAsync ? @"remoteAsync" : @"remote",
    };
    ...
}
```

这些处理完之后,会为JavaScript模块生成一份模块配置表,其中包含模块名、JavaScript方法、常量参数、ModuleID和MethodID:

```
{
  "remoteModuleConfig": {
    ...
    "RCTWebSocketManager": {
      "constants": ...,
      "methods": {
        "connect": {"type":"remote","methodID":0},
        "send":{"type":"remote","methodID":1},
        "close":{"type":"remote","methodID":2}
      },
      "moduleID":11
    },
    ...
  }
}
```

最终,这份配置信息会通过调用[_javaScriptExecutor injectJSONText:asGlobalObjectNamed:callback:]向JavaScript端注册,生成对应的JavaScript模块对象和方法。

通过配置表,我们可以很清晰地将调用的JavaScript方法映射到对应的Native模块方法上了。

5.1.2 通信流程

既然模块配置信息已经生成,那么现在的问题就是如何使用配置信息来完成JavaScript模块与Native模块之间的相互调用,具体如图5-1所示。

JavaScript会维护一个MessageQueue。当调用模块的方法时,调用会被解析为模块名ModuleName、模块方法MethodName、参数Params这三个部分,并将这三个部分传递给MessageQueue,通过前面生成的模块配置remoteModuleConfig解析为ModuleID和MethodID。

MessageQueue是一个很重要的模块,所起的作用与原生代码中的RCTBridge类似,都提供了将模块方法调用与模块通信数据相互转换的功能。同时,MessageQueue提供了4个接口,React Native

就是通过在RCTBridge中调用特定的接口来完成JavaScript与Native之间的通信。

- **processBatch**：获得JavaScript模块的方法调用。
- **invokeCallbackAndReturnFlushedQueue**：调用JavaScript模块执行时传递的回调函数。
- **callFunctionReturnFlushedQueue**：处理Objective-C对JavaScript模块的功能调用。
- **flushedQueue**：用于刷新MessageQueue队列。

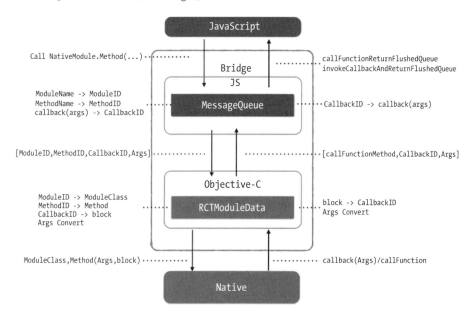

图5-1　JavaScript模块调用Native模块

如果调用参数中包含了回调函数，MessageQueue会将参数Params中的回调函数缓存在本地，并生成一个CallbackID来代替。最终将JavaScript调用以ModuleID、MethodID、Params的形式缓存起来传递给Native：

```
__nativeCall(module, method, params, onFail, onSucc) {
    ...
    onFail && params.push(this._callbackID);
    this._callbacks[this._callbackID++] = onFail;
    onSucc && params.push(this._callbackID);
    this._callbacks[this._callbackID++] = onSucc;
    ...
    this._queue[MODULE_IDS].push(module);
    this._queue[METHOD_IDS].push(method);
    this._queue[PARAMS].push(params);
}
```

这里并不是由JavaScript模块主动将数据传递给Native模块的，而是由Native模块主动调用MessageQueue的接口（processBatch），该接口的返回值为MessageQueue队列中缓存的JavaScript调用的消息数据（[ModuleID, MethodID, Params]）。

Native获得消息数据之后，通过ModuleID索引到对应模块对象RCTModuleData，然后在模块对象中通过MethodID索引到模块方法RCTModuleMethod，这样就定位到调用的JavaScript模块方法所对应的Native模块方法：

```objc
//RCTBatchedBridge.m
- (BOOL)_handleRequestNumber:(NSUInteger)i
                    moduleID:(NSUInteger)moduleID
                    methodID:(NSUInteger)methodID
                      params:(NSArray *)params
{
   ...
   //根据ModuleID映射原生模块
   RCTModuleData *moduleData = _moduleDataByID[moduleID];
   ...
   //根据MethodID映射模块方法
   RCTModuleMethod *method = moduleData.methods[methodID];
   ...
   //转换参数，调用方法
   [method invokeWithBridge:self module:moduleData.instance arguments:params];
      ...
}
```

JavaScript模块与Native模块之间的数据是以JSON类型来传递的，而且Native模块方法的参数必须使用Native的类型。因此，在方法实现之前，需要将JavaScript传递的参数Params根据对应模块方法的参数类型进行处理。在RCTModuleMethod中，框架会通过运行时方法来获取模块方法的参数类型，然后将JSON数据转换为对应的Native类型参数；而对于回调函数CallbackID，则会转换为一个block函数来替换。最后，再由RCTModuleMethod完成方法的调用：

```objc
- (void)processMethodSignature
{
    ...
    _selector = NSSelectorFromString(objCMethodName);
    NSMethodSignature *methodSignature = [_moduleClass instanceMethodSignatureForSelector:_selector];
    NSInvocation *invocation = [NSInvocation invocationWithMethodSignature:methodSignature];
    //获得参数数量
    NSUInteger numberOfArguments = methodSignature.numberOfArguments;

    for (NSUInteger i = 2; i < numberOfArguments; i++) {
        //获得参数类型
        const char *objcType = [methodSignature getArgumentTypeAtIndex:i];
        //参数转换
        ...
    }
    ...
}
```

如果模块方法定义了回调函数，那模块功能代码执行完毕后，需要执行回调函数。再来看看

RCTModuleMethod在参数转换时做了什么处理，回调函数中调用了MessageQueue的接口invokeCallbackAndReturnFlushedQueue，将CallbackID以及参数（需要序列化成JSON数据）传递给MessageQueue，最终由MessageQueue再通过CallbackID找到之前缓存的callback函数，执行JavaScript回调，这样就完成了一次JavaScript模块调用Native模块的流程：

```
//RCTModuleMethod.m
...
[bridge _invokeAndProcessModule:@"BatchedBridge"
                         method:@"invokeCallbackAndReturnFlushedQueue"
                      arguments:@[json, args]];
...
//MessageQueue.js

__invokeCallback(cbID, args) {
  ...
  let callback = this._callbacks[cbID];
  ...
  callback.apply(null, args);
  ...
}
```

在JavaScript端，也会生成一份本地模块配置localModules，Native模块可以通过enqueue-JSCall:args:直接调用JavaScript模块方法，这也是通过传递ModuleID和MethodID来调用的。与前面Native模块中对回调函数的处理相同，通过调用MessageQueue的callFunctionReturnFlushed-Queue接口实现这个流程：

```
/**
 * This method is used to call functions in the JavaScript application context.
 * It is primarily intended for use by modules that require two-way communication
 * with the JavaScript code. Safe to call from any thread.
 */
- (void)enqueueJSCall:(NSString *)moduleDotMethod args:(NSArray *)args;
```

React Native在这个基础上提供了在JavaScript模块中对于Native模块事件的监听处理：在JavaScript端注册事件响应函数，通过Native端发出事件通知，然后JavaScript接受通知后执行对应事件的响应。我们也可以通过这个功能间接地实现对JavaScript模块的调用，具体用法在后面章节中会有介绍。

React Native的通信流程大致就是这样，如果你非常感兴趣，可以继续深入了解相关的源代码，其中的设计思想以及具体实现都非常值得学习。

5.2　自定义Native API组件

在第4章中，我们介绍了React Native目前已经提供了很多基础的API实现，例如网络状态、应用状态、摄像头等功能。而在实际的开发过程中，经常会面临更加复杂的需求，图像识别、数据加密、缓存优化等等，这时React Native提供的API已经无法满足了这些新需求了。在这一节中，我们会告诉大家如何定义自己的API组件。

5.2.1 模块和方法定义

一个普通的Objective-C类以及方法，并不会被系统处理成模块进而被调用。模块必须在编译及运行时向系统注册，同时告诉系统什么属性和方法可以被JavaScript调用。

模块类必须遵循RCTBridgeModule协议。

```
//CalendarManager.h
#import "RCTBridgeModule.h"

@interface CalendarManager : NSObject <RCTBridgeModule>

@end
```

RCTBridgeModule协议中定义了一些模块的基本属性和方法以及一些宏命令。我们可以通过调用对应的宏命令来告诉React Native哪些是我们的模块类，哪些是需要暴露的模块方法：

```
RCT_EXPORT_MODULE //向系统注册模块
RCT_EXPORT_METHOD //暴露模块方法
```

在类实现中，需要调用宏RCT_EXPORT_MODULE(...);来导出模块类。如果需要自定义模块名，在宏命令中传入模块名作为参数即可，如果不填，则默认会使用类名来作为模块名：

```
//CalendarManager.m
@implementation CalendarManager

RCT_EXPORT_MODULE();

@end
```

模块名在JavaScript被映射时，如果模块名前缀包含RCT，会被格式化去除，例如原生模块RCTActionSheetManager在JavaScript中的模块名为ActionSheetManager。模块类完成映射后，会在JavaScript中生成一个模块对象以供调用：

```
var MyModule = require('NativeModules').ModuleName;
```

作为模块方法，同样需要使用宏RCT_EXPORT_METHOD(method)的封装将方法显式导出到JavaScript。宏命令中实现了JavaScript方法与Objective-C方法的映射，会使用Objective-C方法的第一个部分作为JavaScript的方法名。如果需要自定义JavaScript模块的方法，可以使用宏RCT_EXPORT_METHOD(js_name, method)来代替：

```
//Objective-C

RCT_EXPORT_METHOD(doSomething:(NSString *)aString
                  withA:(NSInteger)a
```

```
                    andB:(NSInteger)b)
{
  ...
}

//JavaScript

NativeModules.ModuleName.doSomething(aString,a,b);
```

JavaScript和Objective-C是两个完全不同的语言，所支持的数据类型也各不相同，如果之间需要通信，那必须完成数据类型的转换。React Native中双方的通信数据采用JSON类型来传递，因此支持标准JSON的类型都是支持的：

```
string (NSString)
number (NSInteger, float, double, CGFloat, NSNumber)
boolean (BOOL, NSNumber)
array (NSArray) of any types from this list
map (NSDictionary) with string keys and values of any type from this list
function (RCTResponseSenderBlock)
```

执行Native模块方法前，`RCTModuleMethod`会根据Native方法定义的参数类型通过RCTConvert.h进行转换。在RCTConvert.h中，除了支持JSON标准类型外，也支持一系列常用类型，例如日期类型的数据：

```
//JavaScript
CalendarManager.addEvent('Birthday Party', '4 Privet Drive, Surrey', date.toTime());

//Objective-C
RCT_EXPORT_METHOD(addEvent:(NSString *)name location:
  (NSString *)location date:(NSDate *)date)
{
  //date已经被自动转换为NSDate类型
}
```

这些类型包括但不局限于以下类型（具体可以查看RCTConver.h）：

- `NSDate`、`UIColor`、`UIFont`、`NSURL`、`NSURLRequest`、`UIImage` ...
- `UIColorArray`、`NSNumberArray`、`NSURLArray` ...
- `NSTextAlignment`、`NSUnderlineStyle` ...
- `CGPoint`、`CGSize` ...

5.2.2 回调函数

React Native定义了几种类型的块函数来作为回调函数，`RCTModuleMethod`以及`MessageQueue`会根据不同的类型来作对应的处理。Native中定义的回调函数在执行时都会将数据传递给JavaScript环境，来执行对应的JavaScript函数。

- `(^RCTResponseSenderBlock)(NSArray *)`：接收多个参数的回调函数。

- (^RCTResponseErrorBlock)(NSError *)：接收错误参数的回调函数。
- (^RCTPromiseResolveBlock)(id result)：处理Promise Resolve。
- (^RCTPromiseRejectBlock)(NSError *)：处理Promise Reject。

一般情况下，我们都使用RCTResponseSenderBlock来作为回调函数，此时只需要将一个NSArray对象传入块函数执行即可：

```
RCT_EXPORT_METHOD(findEvents:(RCTResponseSenderBlock)callback)
{
  NSArray *events = ...
  callback(@[[NSNull null], events]);
}
```

如果需要使用回调函数，则要在JavaScript调用模块方法时在对应的参数上定义回调函数，在Native端执行回调时，JavaScript回调函数也将被执行。如果需要使用回调函数中的参数，定义回调函数参数的顺序就要与Native模块中传入的NSArray中的对象顺序保持一致，这样才可以接收到正确的参数。例如，Native模块中回调函数传入的第一个参数是error对象，那么JavaScript回调函数定义的第一个参数将会被赋于这个error对象的JSON值。如果不需要使用回调函数，无需定义回调函数即可，并不会影响正常运行：

```
CalendarManager.findEvents((error, events) => {
  if (error) {
    console.error(error);
  } else {
    this.setState({events: events});
  }
})
```

当然，块函数允许被缓存起来延后执行，例如RCTAlertManager在alertWithArgs:callback:方法中将回调函数callback缓存，当用户点击控件后，在代理事件中调用callback，并释放缓存。当我们这样使用的时候，一定要确保手动缓存的实例会被释放，来防止内存泄漏。示例代码如下：

```
RCT_EXPORT_METHOD(alertWithArgs:(NSDictionary *)args
                  callback:(RCTResponseSenderBlock)callback)
{
  ...
  [_alerts addObject:alertView];
  [_alertCallbacks addObject:callback ?: ^(__unused id unused) {}];
  ...
}

- (void)alertView:(UIAlertView *)alertView clickedButtonAtIndex:(NSInteger)buttonIndex
{
  ...
  RCTResponseSenderBlock callback = _alertCallbacks[index];
  callback(args);
  [_alertCallbacks removeObjectAtIndex:index];
  ...
}
```

如果需要传递错误信息给回调函数，则使用RCTResponseErrorBlock时，方法的参数部分可以直接传入NSError对象。一般来说，当我们希望在JavaScript定义的方法中区分正确返回和错误返回时，会使用RCTResponseErrorBlock回调。当然，也可以像上例中那样，将错误作为第一个参数来表示执行成功与否。因为NSError不是默认支持转换的数据类型，所以不能直接传递NSError对象实例，而需要使用RCTUtil.h的RCTMakeError。

此外，React Native还支持Promise规范的异步编程模式。在支持Promise规范的Native模块方法中，最后两个参数必须强制定义为RCTPromiseResolveBlock和RCTPromiseRejectBlock：

```
RCT_EXPORT_METHOD(doSomethingAsync:(NSString *)aString
                  resolver:(RCTPromiseResolveBlock)resolve
                  rejecter:(RCTPromiseRejectBlock)reject
{
  ...
  if ( error ) {
    reject(error);
  } else {
    resolve(data);
  }
  ...
}
```

调用方法时，可以通过链式执行then方法来处理RCTPromiseResolveBlock的调用，执行catch方法来处理RCTPromiseRejectBlock的调用：

```
NativeModules.ModuleName.doSomethingAsync(aString)
.then(
  (data) => { ... })
.catch(
  (err) => { ... }
);
```

5.2.3 线程

JavaScript代码都是单线程运行的，而在Native模块中，线程问题自然而然会被关注。在React Native中，所有的Native模块都默认运行在各自独立的GCD串行队列上。如果需要特别指定某个线程队列，可以通过- (dispatch_queue_t)methodQueue方法实现：

```
- (dispatch_queue_t)methodQueue
{
  return dispatch_queue_create("com.facebook.React.NameQueue",
    DISPATCH_QUEUE_SERIAL);
}
```

模块中所有的模块方法都会运行在同一线程队列中，如果某些方法需要单独指定队列，可以使用dispatch_async：

```
RCT_EXPORT_METHOD(doSomethingExpensive:(NSString *)param
```

```objc
    callback:(RCTResponseSenderBlock)callback)
{
  dispatch_async(dispatch_get_global_queue(
    DISPATCH_QUEUE_PRIORITY_DEFAULT, 0), ^{
    //Call long-running code on background thread
    ...
    //You can invoke callback from any thread/queue
    callback(@[...]);
  });
}
```

如果多个模块需要共享一个线程队列，那么我们必须手动缓存共享的队列实例，并在 methodQueue 中返回共享实例，而不是创建一个相同标签的实例。

5.2.4 常量导出

React Native 还支持 Native 模块暴露一些常量数据供 JavaScript 模块方法使用，主要有以下用法。

- Native 组件中的常量值，例如版本和事件名称等。
- Native 中定义枚举在 JavaScript 中使用的对应值。
- 边界定义，例如控件允许的最小尺寸或者默认尺寸等。

可以定义一些默认值以及一些参数的枚举值来提高使用 JavaScript 时的可读性。

通过实现 constantsToExport 方法，可以返回一个字典对象：

```objc
- (NSDictionary *)constantsToExport
{
  return @{ @"firstDayOfTheWeek": @"Monday" };
}
```

在 JavaScript 中，可以直接访问字典的 key 值来访问字典对象：

```
//Monday
CalendarManager.firstDayOfTheWeek
```

常量会在框架初始化期间被导出到模块配置表中，并在生成 JavaScript 模块对象时并入该对象中，因此，在运行时对 constantsToExport 的字典的任何修改将无法生效。

上面提到，可以在常量导出中将枚举类型 NS_ENUM 导出以供 JavaScript 作为参数来使用。但是问题来了，通过常量导出枚举变量时，枚举的值都是以整型数字存储的。那么，在 JavaScript 中调用模块方法，将常量导出中的枚举值作为参数传递到 Native 时，对于 React Native 而言，我们只是单纯地传递了一个整型参数。因此，我们需要为自定义的枚举类型进行 RCTConvert 扩展，这样在模块方法调用中使用常量导出的枚举值，通信到 Native 中时，会从整型自动转换为定义的枚举类型。

定义一个枚举：

```objc
typedef NS_ENUM(NSInteger, UIStatusBarAnimation) {
  UIStatusBarAnimationNone,
```

```
UIStatusBarAnimationFade,
UIStatusBarAnimationSlide,
};
```

实现RCTConvert对于UIStatusBarAnimation类型的扩展：

```
@implementation RCTConvert (StatusBarAnimation)
  RCT_ENUM_CONVERTER(
    UIStatusBarAnimation,
    (@{ @"statusBarAnimationNone": @(UIStatusBarAnimationNone),
        @"statusBarAnimationFade": @(UIStatusBarAnimationFade),
        @"statusBarAnimationSlide" : @(UIStatusBarAnimationSlide),
       UIStatusBarAnimationNone, integerValue)

@end
```

然后在常量导出中会返回对应的枚举值：

```
- (NSDictionary *)constantsToExport
{
  return @{ @"statusBarAnimationNone" : @(UIStatusBarAnimationNone),
            @"statusBarAnimationFade" : @(UIStatusBarAnimationFade),
            @"statusBarAnimationSlide" : @(UIStatusBarAnimationSlide) }
};
```

在JavaScript方法调用中，只要将枚举类型的参数位置传入模块常量导出中对应的值，例如Module.statusBarAnimationSlide，这个参数在Native模块中就会被自动转换为对应的枚举类型UIStatusBarAnimation：

```
//Native
RCT_EXPORT_METHOD(updateStatusBarAnimation:(UIStatusBarAnimation)animation
                                completion:(RCTResponseSenderBlock)callback);

//JavaScript
Module.updateStatusBarAnimation(Module.statusBarAnimationSlide, callback
);
```

5.2.5 事件

React Native在Native向JavaScript传递消息机制的基础上实现了一个非常低耦合的消息事件订阅系统，Native通过RCTEventDispatcher向JavaScript端的EventEmitter模块发送事件消息，由EventEmitter模块通知该事件的订阅者来执行事件的响应。在大多数场景下，只需要使用这种通知的方式间接完成Native对JavaScript的调用。

首先，在JavaScript端对事件进行订阅，并且添加事件响应函数：

```
var { NativeAppEventEmitter } = require('react-native');

var subscription = NativeAppEventEmitter.addListener(
```

```
'EventReminder',
(reminder) => console.log(reminder.name)
);
```

当在Native模块上发出事件通知时,EventEmitter模块则会执行所有注册EventReminder事件的响应函数:

```
//引入RCTEventDispatcher头文件
//#import "RCTEventDispatcher.h"

[self.bridge.eventDispatcher sendAppEventWithName:@"EventReminder"
                                             body:@{@"name": eventName}];
```

React Native定义了不同的接口以及接收者来区分事件的类型:

```
//发送应用相关的事件,例如数据更新
//NativeAppEventEmitter
- (void)sendAppEventWithName:(NSString *)name body:(id)body;

//发送设备相关的事件,例如地理定位和屏幕旋转
//DeviceEventEmitter
- (void)sendDeviceEventWithName:(NSString *)name body:(id)body;
```

最后,我们需要在合适的时候手动取消事件的订阅subscription.remove();,否则这个订阅可能会导致内存泄漏。

5.2.6 实战

前面我们简单介绍了如何创建Native模块,现在创建一个简单的Native模块,作为对已有功能的补充。通过该模块获得当前的屏幕尺寸Dimensions,并且提供注册事件监控屏幕方向变化。

我们在项目工程中创建Native模块类RCTDeviceExtension,它继承RCTBridgeModule协议:

```
#import "RCTBridgeModule.h"

@interface RCTDeviceExtension : NSObject<RCTBridgeModule>

@end
```

然后在RCTDeviceExtension.m中将该类注册为Native模块类:

```
@implementation RCTDeviceExtension

RCT_EXPORT_MODULE();

@end

//在JavaScript代码中查询原生模块
console.log(require('NativeModules'));
```

运行上述代码,得到的效果如图5-2所示,从中可以看到RCTDeviceExtension模块已经被加载

到模块配置中了，同时在JavaScript模块中生成了DeviceExtension对象。

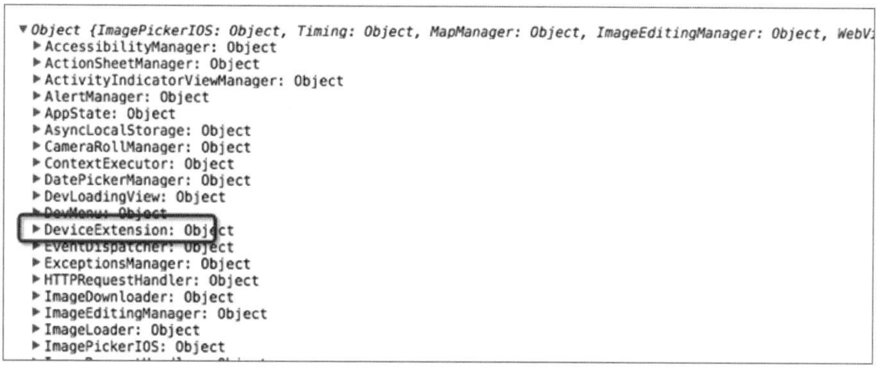

```
▼Object {ImagePickerIOS: Object, Timing: Object, MapManager: Object, ImageEditingManager: Object, WebVi
  ▶AccessibilityManager: Object
  ▶ActionSheetManager: Object
  ▶ActivityIndicatorViewManager: Object
  ▶AlertManager: Object
  ▶AppState: Object
  ▶AsyncLocalStorage: Object
  ▶CameraRollManager: Object
  ▶ContextExecutor: Object
  ▶DatePickerManager: Object
  ▶DevLoadingView: Object
  ▶DevMenu: Object
  ▶DeviceExtension: Object
  ▶EventDispatcher: Object
  ▶ExceptionsManager: Object
  ▶HTTPRequestHandler: Object
  ▶ImageDownloader: Object
  ▶ImageEditingManager: Object
  ▶ImageLoader: Object
  ▶ImagePickerIOS: Object
```

图5-2　加载模块

接着开始实现模块方法。添加一个方法getDynamicDimensions()来返回当前的屏幕尺寸，并且注册一个模块方法getDynamicDimensions:(RCTResponseSenderBlock)callback来返回我们需要的结果：

```
static NSDictionary *DynamicDimensions() {
    //提供当前屏幕的尺寸
    CGFloat width = MIN(RCTScreenSize().width, RCTScreenSize().height);
    CGFloat height = MAX(RCTScreenSize().width, RCTScreenSize().height);
    CGFloat scale = RCTScreenScale();
    if (UIDeviceOrientationIsLandscape([UIDevice currentDevice].orientation)) {
      width = MAX(RCTScreenSize().width, RCTScreenSize().height);
      height = MIN(RCTScreenSize().width, RCTScreenSize().height);
    }
    return @{@"width":@(width),
             @"height":@(height),
             @"scale":@(scale)};
}

RCT_EXPORT_METHOD(getDynamicDimensions:(RCTResponseSenderBlock)callback) {
    callback(@[[NSNull null], DynamicDimensions()]);
}
```

要实现屏幕方向的监听，需要在模块初始化时向消息中心注册一个屏幕方向变化的系统事件UIDeviceOrientationDidChangeNotification：

```
- (instancetype)init
{
  self = [super init];
  if (self) {
    [[NSNotificationCenter defaultCenter] addObserver:self
      selector:@selector(orientationDidChange:)
      name:UIDeviceOrientationDidChangeNotification object:nil];
  }
```

```objc
    return self;
}

- (void)dealloc
{
    [[NSNotificationCenter defaultCenter] removeObserver:self];
}
```

所有的Native模块都在框架初始化时创建一个实例，由React Native统一管理，并不存在多实例事件被重复注册的问题。在系统事件回调中，通过RCTEventDispatcher将事件orientationDidChange通知到JavaScript。这里必须先要引入RCTEventDispatcher头文件，否则调用时Xcode会报错：

```objc
@synthesize bridge = _bridge;

- (void)orientationDidChange:(id)noti
{
  [_bridge.eventDispatcher sendDeviceEventWithName:
    @"orientationDidChange" body:
    @{    @"Orientation":UIDeviceOrientationIsLandscape(
    [UIDevice currentDevice].orientation) ?@"Landscape":@"Portrait",
    @"Dimensions":DynamicDimensions()}];
}
```

最后将事件变量作为常量导出，供JavaScript注册事件时使用：

```objc
- (NSDictionary *)constantsToExport
{
    return @{@"EVENT_ORIENTATION":@"orientationDidChange"};
}
```

现在，在JavaScript中调用我们的模块来看一下结果。

调用getDynamicDimensions方法获得当前屏幕的尺寸：

```javascript
var DeviceExtension = require('NativeModules').DeviceExtension;

DeviceExtension.getDynamicDimensions((error,dimensions) => {
  console.log(dimensions);
});


//Console Log

Object {width: 375, scale: 2, height: 667}
```

然后接收屏幕旋转的事件：

```javascript
var RCTDeviceEventEmitter = require('RCTDeviceEventEmitter');

var subscription =
  RCTDeviceEventEmitter.addListener('orientationDidChange',
```

```
(dimensions) => {
//Object {Orientation: "Landscape", Dimensions: Object}
//Object {Orientation: "Portrait", Dimensions: Object}
console.log(dimensions);
});

//subscription.remove();
```

这样一个Native模块就完成了。不难发现，React Native的确提供了一个非常便捷的方式来扩展Native模块。如果要把模块做成第三方组件的话，还有一些工作要做：首先以一个静态库工程来编译模块代码，提供JavaScript的封装，最后创建Package.json来支持node的引用。

在项目目录node_modules/react-native-device-extension下新建一个静态库项目RCTDeviceExtension，并且加入模块类文件RCTDeviceExtension，如图5-3所示。

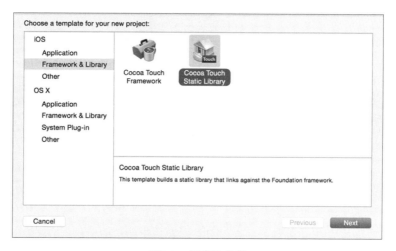

图5-3　新建静态库

然后在该目录下新建RCTDeviceExtension.ios.js与package.json。

```
node_modules
    |--react-native-device-extension
        |- package.json
        |- RCTDeviceExtension.ios.js
        |- RCTDeviceExtension.xcodeproj
        |- RCTDeviceExtension.h
        |- RCTDeviceExtension.m
```

在应用工程中引入模块的静态库，在Libraries加入静态库的xcodeproj文件，并在Build Phase → Link Binary With Libraries中引入模块的静态库文件，如图5-4所示。

```
  ┌─────────────────────────────────────────────────┐
  │ ▼ Linked Frameworks and Libraries               │
  │   Name                                  Status  │
  │   📦 libRCTDeviceExtension.a            Required│
  │   📦 libReact.a                         Required│
  │   📦 libRCTActionSheet.a                Required│
  │   📦 libRCTGeolocation.a                Required│
  │   📦 libRCTImage.a                      Required│
  │   📦 libRCTLinking.a                    Required│
  │   📦 libRCTNetwork.a                    Required│
  │   📦 libRCTSettings.a                   Required│
  │   📦 libRCTText.a                       Required│
  │   📦 libRCTVibration.a                  Required│
  │   📦 libRCTWebSocket.a                  Required│
  │   + −                                           │
  └─────────────────────────────────────────────────┘
```

图5-4 引入静态库文件

这时会发现错误'RCTBridgeModule.h' file not found，这是因为静态库需要外部引用React的基础库，需要添加相关头文件的路径。此时在项目中Build Settings→Search Paths→Header Search Paths增加$(SRCROOT)/../react-native/React，并且把模式修改为recursive即可。需要说明的是，这里的路径取决于RCTDeviceExtension.xcodeproj与React.xcodeproj的相对路径。

然后对JavaScript模块进行封装。对于JavaScript人员来说，有可能不是很熟悉Native的代码，因此提供对应的JavaScript封装并且辅以注释还是很有必要的：

```javascript
//RCTDeviceExtension.ios.js

'use strict'

var RCTDeviceEventEmitter = require('RCTDeviceEventEmitter');
var ExtensionModule = require('NativeModules').DeviceExtension;

var __deviceSubscriptions = {};

var DeviceExtension = {
  //事件列表
  events : {
    EVENT_ORIENTATION: ExtensionModule.EVENT_ORIENTATION
  },
  //获得当前屏幕大小
  getDynamicDimensions: function(handler: Function) {
    ExtensionModule.getDynamicDimensions(handler);
  },
  //监听事件
  addLisener: function(event: String, handler: Function) {
    if (event == self.events.EVENT_ORIENTATION) {
      __deviceSubscriptions[handler] = RCTDeviceEventEmitter.addListener(event, handler);
    }
  },
  //移除事件监听
  removeLisener: function(event: String, handler: Function) {
```

```
      if (!__deviceSubscriptions[handler]) {
        return;
      }
      __deviceSubscriptions[handler].remove();
      __deviceSubscriptions[handler] = null;
  }
}

module.exports = DeviceExtension;
```

最后,在目录下添加package.json文件。package.json文件的内容如下:

```
//package.json
...
{
  "name": "react-native-device-extension",
  "version": "0.1",
  "main": "./RCTDeviceExtension.ios.js"
}
...
```

然后我们的模块就可以与官方模块采用相同的方式引入:

```
//index.ios.js

var DeviceManager = require('react-native-device-extension');

var RNDemo = React.createClass({

  getInitialState: function() {
    returun {
      orientation: 'unkown'
    };
  },

  viewDidOrientation: function(event) {
    this.state.orientation = event.Orientation;
  },

  componentDidMount: funcion() {
    DeviceManager.addLisener(DeviceManager.events.
      DEVICE_ORIENTATION_EVENT, this.viewDidOrientation);
  },

  componentWillUnmount: function() {
    DeviceManager.removeLisener(DeviceManager.events.
      DEVICE_ORIENTATION_EVENT, this.viewDidOrientation);
  },

  ...

});
```

5.3 构建 Native UI 组件

React Native提供了许多基础组件的封装，例如Navigator、MapView、DatePickerIOS、ScrollView等。这些基础组件结合使用React，也可以构建出一些复合组件。当然，因为设计的原因，某些包含大量子控件的复合组件（例如表格类组件）在性能上会有不足。原生组件经过长时间的发展，已经有了非常不错的积累，许多优秀的UI组件被开源出来供大家选择。React Native也提供了便捷的方式，将原生组件抽象成ReactJS中的组件对象来供JavaScript端调用。

5.3.1 概述

扩展的Native UI组件也是一个Native模块。相对API组件来说，UI组件还需要被抽象出供React使用的标签，提供标签属性，响应用户行为等。在React中创建UI组件时，都会生成reactTag来作为唯一标识。JavaScript UI与原生UI都将通过reactTag进行关联。JavaScript UI的更新会通过调用RCTUIManager模块的方法来映射成原生UI的更新。当原生UI被通知改变时，会通过reactTag来定位UI实例来进行更新操作，所有的UI更新操作并不会马上执行，而是会被缓存在一个UIBlocks中，每次通信完毕，再由主线程统一执行UIBlocks中的更新。在帧级别的通信频率下，让原生UI无缝地响应JavaScript的改变。

5.3.2 UI 组件的定义

要构建Native UI组件，先要创建UI的管理类，这个管理类需要继承RCTViewManager类。实质上，RCTViewManager类同样遵循RCTBridgeModule的协议。然后再实现- (UIView *)view接口，在接口返回Native UI的实例：

```objc
#import "RCTViewManager.h"

@interface RCTMapManager : RCTViewManager
@end

@implementation RCTMapManager

RCT_EXPORT_MODULE()

- (UIView *)view
{
  return [[MKMapView alloc] init];
}

@end
```

需要注意的是，UI组件的样式都是由JavaScript来控制的，因此在- (UIView *)view中实例化时，任何对于样式的操作都会被JavaScript模块中对应的样式覆盖。同时值得注意的是，在实例化时一般不对View的Frame进行设置。如果UI组件内部的UI或者图层不支持自适应，则需要在UI组件的- (void)layoutSubviews方法中对内部控件实现自适应布局。

沿用系统的命名规范，扩展的UI组件模块都以Manager为后缀，在React中使用时只需要在JavaScript中导出对应的原生组件对象即可。组件名需要过滤类名后缀Manager，所有的组件对象导出后都可以通过组件标签引用：

```
...
var { requireNativeComponent } = require('react-native');
module.exports = requireNativeComponent('RCTMap', null);

//可以在render方法中这样使用<RCMap/>

...
```

5.3.3　UI 组件属性

原生组件的属性也需要桥接到JavaScript，以标签属性的形式访问。

一般情况下，我们使用RCT_EXPORT_VIEW_PROPERTY来桥接Native UI的属性：

```
RCT_EXPORT_VIEW_PROPERTY(pitchEnabled, BOOL)
```

默认情况下，JavaScript标签属性与Native属性相同。如果需要另外定义，可以使用RCT_REMAP_VIEW_PROPERTY。

然后就可以在JavaScript中为组件标签设置属性了：

```
<RCTMap pitchEnabled={false} />
```

在RCTViewManager.m中可以看到，React Native对于一些默认的View属性自动做了桥接View，我们可以直接使用：

```
//RCTViewManager.m
...
RCT_EXPORT_VIEW_PROPERTY(backgroundColor, UIColor)
RCT_REMAP_VIEW_PROPERTY(accessible, isAccessibilityElement, BOOL)
...
```

与模块方法相同，属性的类型支持标准JSON对象。同时，RCTConvert.h中也提供一些常用类型的自动转换，例如UIColor和NSDate。如果是默认不支持的类型属性，可以通过宏RCT_CUSTOM_VIEW_PROPERTY(name, type, viewClass)来完成自定义类型属性的扩展，其中参数name为属性名，type为属性的类型，viewClass则是Native组件的类型。在宏命令后的方法实现中，对控件的对应属性进行赋值。示例如下：

```
//RCTMapManager.m
RCT_CUSTOM_VIEW_PROPERTY(region, MKCoordinateRegion, RCTMap)
{
```

```
  [view setRegion:json ? [RCTConvert MKCoordinateRegion:json] :
    defaultView.region animated:YES];
}
```

我们可以为视图的属性类型创建自定义的RCTConvert扩展来进行类型转换。以下是对MKCoordinateRegion类型的RCTConvert扩展：

```
@implementation RCTConvert(CoreLocation)

RCT_CONVERTER(CLLocationDegrees, CLLocationDegrees, doubleValue);
RCT_CONVERTER(CLLocationDistance, CLLocationDistance, doubleValue);

+ (CLLocationCoordinate2D)CLLocationCoordinate2D:(id)json
{
  json = [self NSDictionary:json];
  return (CLLocationCoordinate2D){
    [self CLLocationDegrees:json[@"latitude"]],
    [self CLLocationDegrees:json[@"longitude"]]
  };
}
@end

@implementation RCTConvert(MapKit)

+ (MKCoordinateSpan)MKCoordinateSpan:(id)json
{
  json = [self NSDictionary:json];
  return (MKCoordinateSpan){
    [self CLLocationDegrees:json[@"latitudeDelta"]],
    [self CLLocationDegrees:json[@"longitudeDelta"]]
  };
}

+ (MKCoordinateRegion)MKCoordinateRegion:(id)json
{
  return (MKCoordinateRegion){
    [self CLLocationCoordinate2D:json], //转换
    [self MKCoordinateSpan:json]
  };
}
```

此外，我们也可以采用与API组件相同的方式，结合常量导出与RCTConvert扩展提供枚举类型的支持：

```
@implementation RCTConvert(UIAccessibilityTraits)

RCT_MULTI_ENUM_CONVERTER(UIAccessibilityTraits, (@{
  @"none": @(UIAccessibilityTraitNone),
  ...
  @"pageTurn": @(UIAccessibilityTraitCausesPageTurn),
}), UIAccessibilityTraitNone, unsignedLongLongValue)

@end
```

```
//RCTViewManager.m
...
RCT_EXPORT_VIEW_PROPERTY(accessibilityTraits, UIAccessibilityTraits)
...
```

5.3.4 组件方法

Native UI组件同样支持模块方法，其方法定义中必须包含由JavaScript传递过来的reactTag，其实现逻辑需要封装在RCTUIManager的addUIBlock接口的块函数中执行。这样，在块函数中，我们可以使用RCTUIManager维护的ViewRegistry通过reactTag获得调用方法的组件实例：

```
RCT_EXPORT_METHOD(reload:(NSNumber *)reactTag) {
  [self.bridge.uiManager addUIBlock:^(__unused RCTUIManager *uiManager, RCTSparseArray *viewRegistry)
  {
    id view = viewRegistry[reactTag];
    //完成对控件重载方法的调用
  }];
}
```

在JavaScript中，需要为组件设置引用ref，调用方法时通过引用React.findNodeHandle(ref)来获得组件的reactTag，然后将其作为组件模块方法对应的参数传入：

```
var RCT_UI_REF = ...;

<RCTCustomUI
  ref={RCT_UI_REF}
  ...
/>

//方法调用
RCTCustomUIManager.reload(
  React.findNodeHandle(this.refs[RCT_UI_REF])
);
```

5.3.5 事件

大多数控件都会接受用户的行为，例如TextField需要接受用户输入文本和点击按钮的行为，图表控件会接受用户的手势等，这时我们需要控件能够针对用户的行为作出响应。在纯Native的编码中，这很容易做到。例如为控件增加UIControlEvent的监听，实现对应的Delegate方法等。但是在React Native框架中，我们还需要把这个事件通知到JavaScript，最后由JavaScript端来完成事件的响应，具体步骤如下所示。

首先，设置事件响应函数：

```
- (void)init {
  [control addTarget:self action:@selector(controlValueChanged:)
    forControlEvents:UIControlEventValueChanged];
}
```

```objc
- (void)controlValueChanged:(id)sender {
    //处理控件的事件
}
```

然后设置Delegate回调函数:

```objc
- (void)tableView:(UITableView *)tableView didSelectRowAtIndexPath:(NSIndexPath *)indexPath
{
    //处理Table点击事件
}
```

在React Native中,还要在原生控件响应用户事件的地方,通过事件派发器RCTEvent-Dispatcher的sendInputEventWithName方法来将事件发送给JavaScript模块。在React Native中,事件名在Native模块中会进行格式化处理,例如在Native端,change、onChange和TopChange 都会格式化为topChange事件,而JavaScript都会以onChange事件属性来响应:

```objc
- (void)controlValueChanged:(id)sender {
    NSDictionary *event = @{
        @"target": sender.reactTag,
        @"value": @(sender.value),
    };
    [self.bridge.eventDispatcher sendInputEventWithName:@"topChange" body:event];
}
```

最后,由JavaScript组件响应函数处理:

```jsx
<RCTCustomUI
  onChange={changeHandler}
  ...
/>
```

在RCTViewManager中,默认定义了一些事件,这些事件会自动与JavaScript标签中的onEventName属性进行绑定,具体如下:

- press
- change
- focus
- blur
- submitEditing
- endEditing
- touchStart
- touchMove
- touchCancel
- touchEnd

如果需要绑定自定义事件,可以通过在Manager模块类中重写- (NSArray *)customBubbling-EventTypes接口来实现:

```
- (NSArray *)customBubblingEventTypes
{
    return @[
        @"downloadComplete"
    ];
}
```

此时在标签中就可以使用标签属性调用响应函数了：

```
<RCTCustomUI
  onDownloadComplete ={downloadHandler}
  ...
/>
```

发送事件时，也会将一些数据通过sendInputEventWithName:body:接口的body参数传递给JavaScript，此时数据需要封装为一个字典类型，必须包含target字段来存放reactTag，让JavaScript能够通知到对应的组件来执行响应函数。在JavaScript中，可以通过响应函数的参数body.nativeEvent直接访问字典中的数据：

```
var changeHandler = function(body) {
    console.log(body.nativeEvent.userData);
}
```

5.3.6 实例

在这一节中，我们为大家演示如何将一个原生组件扩展为React Native的UI组件。这里使用一个开源的PieChart组件，它在GitHub上的代码仓库地址为https://github.com/xyfeng/XYPieChart。

这里我们直接以node模块的形式来创建UI组件扩展，具体步骤如下。

(1) 在node_modules/react-native-xypiechart下新建模块目录。
(2) 创建静态库工程XYPieChart，并且引入开源组件XYPieChart类，同时创建XYPieChart-Manager模块管理类。
(3) 添加PieChart.ios.js和package.json文件。
(4) 为应用工程引入扩展静态库libXYPieChart.a。

相对API组件来说，UI组件更加复杂，因为其中包含了许多自定义的属性和方法。一般来说，组件都会使用JavaScript通过React进行封装，并添加适当的注释来提高可读性。

node模块的目录结构如下：

```
node_modules
    |--react-native-xypiechart
        |- package.json
        |- PieChart.ios.js
        |- XYPieChart.h
        |- XYPieChart.m
        |- XYPieChart.xcodeproj
        |- XYPieChartManager.m
        |- XYPieChartManager.m
```

首先，XYPieChartManager需要继承RCTViewManager类，同时实现中需要导出UI模块，并且实现-(UIView *)view来返回UI的实例：

```objc
//XYPieChartManager.h
@interface XYPieChartManager : RCTViewManager

@end

//XYPieChartManager.m
@implementation XYPieChartManager

RCT_EXPORT_MODULE()

- (UIView *)view
{
  XYPieChart *pieChart = [[XYPieChart alloc] init];
  pieChart.delegate = self;
  return pieChart;
}

@end
```

接着在PieChart.ios.js中直接导出XYPieChart组件，将在后续章节中完成ReactJS的封装：

```js
//PieChart.ios.js
var { requireNativeComponent, } = require('react-native');
var XYPieChart = requireNativeComponent('XYPieChart', null);

module.exports = XYPieChart;
```

然后直接使用组件标签来构建视图：

```js
//index.ios.js
...
var PieChart = require('react-native-xypiechart');
var RNDemo = React.createClass({
  ...
  render: function() {
    return (
      <View style={styles.container}>
        <PieChart style={styles.chart} />
      </View>
    )
  }
});

var styles = StyleSheet.create({
  ...
  chart: {
    width:200,
    height:200,
    borderWidth:1
  }
```

});
...

运行上述代码,可以看到组件已经被渲染到屏幕中了,如图5-5所示。

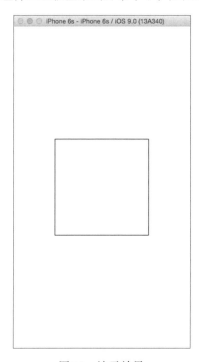

图5-5 演示效果

接着为组件扩展标签属性,这可以根据原生控件的属性选择性添加:

```
//XYPieChart.h
...

@property(nonatomic, weak) id<XYPieChartDataSource> dataSource;
@property(nonatomic, weak) id<XYPieChartDelegate> delegate;
...
@property(nonatomic, strong) UIFont    *labelFont;    //饼图显示标签字体
@property(nonatomic, strong) UIColor   *labelColor;   //饼图显示标签字体颜色
@property(nonatomic, assign) BOOL      showPercentage; //显示百分比还是原值
...
```

可以看出,控件需要通过DataSource代理来为Chart设置数据。如果将Manager类设置为数据源实现DataSource方法,随之而来的问题是,并没有特别好的方案在视图释放时手动释放缓存数据。这里推荐一个简洁的办法,通过扩展为控件封装一个Data属性,由控件自身来实现数据源实现:

```objc
@interface XYPieChart (ReactCategory)<XYPieChartDataSource>

//{'label':'',value:'',color:''}
@property (nonatomic, strong) NSArray *chartData;
@end

@implementation XYPieChart (ReactCategory)

- (NSArray *)chartData
{
  return (NSArray *)objc_getAssociatedObject(self, @selector(chartData));
}

- (void)setChartData:(NSArray *)chartData
{
  objc_setAssociatedObject(self, @selector(chartData), chartData,
    OBJC_ASSOCIATION_RETAIN_NONATOMIC);
  [self reloadData];
}

- (NSUInteger)numberOfSlicesInPieChart:(XYPieChart *)pieChart
{
  return [self.chartData count];
}

- (CGFloat)pieChart:(XYPieChart *)pieChart valueForSliceAtIndex:(NSUInteger)index
{
  return [self.chartData[index][@"value"] intValue];
}

- (UIColor *)pieChart:(XYPieChart *)pieChart colorForSliceAtIndex:(NSUInteger)index
{
  return [RCTConvert UIColor:self.chartData[index][@"color"]];
}

- (NSString *)pieChart:(XYPieChart *)pieChart textForSliceAtIndex:(NSUInteger)index
{
  return self.chartData[index][@"label"];
}

@end
```

这样我们只需要为控件扩展chartData属性,设置数据显示:

```objc
//XYPieChartManager.m

RCT_EXPORT_VIEW_PROPERTY(chartData, NSArray)

RCT_EXPORT_VIEW_PROPERTY(showPercentage, BOOL)

RCT_EXPORT_VIEW_PROPERTY(labelFont, UIFont)

RCT_EXPORT_VIEW_PROPERTY(labelColor, UIColor)
```

```
//index.ios.js

<PieChart style={styles.chart}
  chartData={chartData}
  showPercentage={false} />
```

现在来看一下控件效果，如图5-6所示。

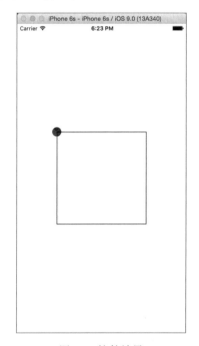

图5-6　控件效果

可以看到，饼图并没有如期望的那样展现。查看XYPieChart的源代码，可以发现控件并不支持自适应。在初始化时，所有的图形图层都根据当时的尺寸创建完毕了。因此，我们必须对此进行修改，以便在布局时UI的大小发生变化后重绘图形，这样就可以通过JavaScript的样式来操控扩展的UI组件了：

```
- (void)layoutSubviews
{
  [super layoutSubviews];
  if (!CGRectEqualToRect(self.bounds, _pieView.bounds)) {
    [self initPieView];
    [self reloadData];
  }
}

- (void)initPieView
{
```

//重建图层
}

运行示例,可以看到饼图已经正常显示在屏幕中了,如图5-7所示。

接着尝试修改一下标签属性showPercentage={true},此时再运行代码,发现饼图中的文字换成了百分比显示,如图5-8所示。

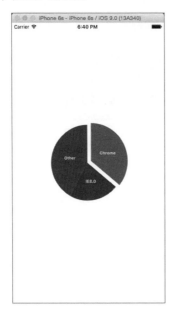
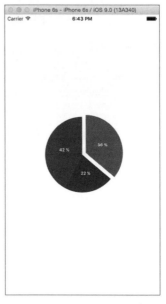

图5-7　饼图　　　　图5-8　修改标签属性后的饼图

我们希望饼图中某一部分被点击的时候可以通知JavaScript来处理相关的响应。先在XYPieChartManager中实现饼图被选中的代理方法pieChart:didSelectSliceAtIndex:,在代理方法中通过事件派发器来向JavaScript端发送用户输入事件change,并将选中饼图的数据包装成一个字典对象通过body参数传递,字典对象中必须定义target的键值来存储组件的reactTag,否则运行时将会出错,无法通知到JavaScript中的组件对象。在JavaScript中,可以通过使用onChange标签来设置响应函数来处理点击事件:

```
//XYPieChartManager.m
- (void)pieChart:(XYPieChart *)pieChart didSelectSliceAtIndex:(NSUInteger)index
{
  NSDictionary *event = @{
    @"target": [pieChart reactTag],
    @"data" : pieChart.chartData[index]
  };
  [self.bridge.eventDispatcher sendInputEventWithName:@"change" body:event];
}
```

//index.ios.js

```
selectHandler: function(body) {
  //body.NativeEvent.data
}
...
<PieChart onChange={(this.selectHandler)} ... />
```

当点击图表中某一块时，JavaScript中定义的selectHandler就会被执行，在函数中可以将被选中区块的数据展示出来或者做一些其他操作。

最后，我们还需要将其封装成第三方组件，具体操作与API组件一样，这里就不重复了。

React Native提供了非常便捷的扩展方法来将原生组件抽象到JavaScript中使用，这对于框架本身的发展非常有利。目前，国外已经有一部分应用尝试使用React Native，详见http://facebook.github.io/react-native/showcase.html。我们相信，很快就会有更多的React Native应用进入大家的视线。

第6章 组件封装

在第3章和第4章中，我们分别学习了React Native的常用组件和API。我们也希望将业务拆分成组件，这样有利于团队合作和代码的后期维护。其实，第3章涉及了业务组件的封装，第5章涉及了Native组件的扩展和封装，这一章我们将系统介绍一些业务组件的封装。一款App由很多组件组成，因此，合理地分解App的功能需要大量的实践和探索。合理地划分组件将会极大地推动团队合作，减少开发成本。

6.1 二级菜单组件

开发组件的好处有很多，最为明显的是复用和独立功能模块。在搜索功能模块中，经常会遇到二级菜单，尤其是在LBS相关的应用中。图6-1是我们常常需要的功能需求，这个需求就是二级菜单。

图6-1 二级菜单

在这一节中，我们就是要一步步实现该组件。

6.1.1 静态组件的实现

开发组件之前，我们需要将一个大组件划分成颗粒度适中的组件。这里我们以二级菜单为例进行介绍。该组件复杂的地方主要在于需要切换不同视图，以期出现菜单联动效果。

动态组件是指可以从外部传入数据和属性的组件，可以复用。静态组件是不具有重用特征的组件，由静态数据组成。静态组件的实现相对来说比较简单，只要开发页面即可，不要考虑属性数据的传递，具体的代码如下：

```
var React = require('react-native');
var {
  AppRegistry,
  StyleSheet,
  Text,
  View,
  ScrollView,
} = React;

var MenuList = React.createClass({
  getInitialState: function(){
    return {
      wholeArea: false,
      hotBusiness: true,
      hotDistrict: false,
      wholeAreaFFF:{},
      hotBusinessFFF:{backgroundColor:'#fff'},
      hotDistrictFFF:{}
    };
  },
  render: function(){
    return (
      <View style={styles.container}>
        <View style={[styles.row, styles.header]}>
          <View style={[styles.flex_1, styles.center]}>
            <Text style={[styles.header_text, styles.active_blue]}>全部区域</Text>
          </View>
          <View style={[styles.flex_1, styles.center]}>
            <Text style={[styles.header_text]}>地铁沿线</Text>
          </View>
        </View>
        <View style={[styles.row, styles.flex_1]}>
          <ScrollView style={[styles.flex_1, styles.left_pannel]}>
            <Text onPress={this.wholeArea} style={[styles.left_row,
              this.state.wholeAreaFFF]}> 全部区域</Text>
            <Text onPress={this.hotBusiness} style={[styles.left_row,
              this.state.hotBusinessFFF]}> 热门商圈</Text>
            <Text onPress={this.hotDistrict} style={[styles.left_row,
              this.state.hotDistrictFFF]}> 热门行政区</Text>
          </ScrollView>

          {
```

```
                this.state.wholeArea ?
                <ScrollView style={[styles.flex_1, styles.right_pannel]}>
                  <Text style={styles.left_row}>全部区域</Text>
                </ScrollView>
                : null
              }

              {
                this.state.hotBusiness ?
                <ScrollView style={[styles.flex_1, styles.right_pannel]}>
                  <Text onPress={this.} style={styles.left_row}>虹桥地区</Text>
                  <Text style={styles.left_row}>徐家汇地区</Text>
                  <Text style={styles.left_row}>淮海路商业区</Text>
                  <Text style={styles.left_row}>静安寺地区</Text>
                  <Text style={styles.left_row}>上海火车站地区</Text>
                  <Text style={styles.left_row}>浦东陆家嘴金融贸易区</Text>
                  <Text style={styles.left_row}>四川北路商业区</Text>
                  <Text style={styles.left_row}>人民广场地区</Text>
                  <Text style={styles.left_row}>南翔、安亭汽车城</Text>
                </ScrollView>
                : null
              }

              {
                this.state.hotDistrict ?
                <ScrollView style={[styles.flex_1, styles.right_pannel]}>
                  <Text style={styles.left_row}>静安区</Text>
                  <Text style={styles.left_row}>徐汇区</Text>
                  <Text style={styles.left_row}>长宁区</Text>
                  <Text style={styles.left_row}>黄浦区</Text>
                  <Text style={styles.left_row}>虹口区</Text>
                  <Text style={styles.left_row}>宝山区</Text>
                  <Text style={styles.left_row}>闸北区</Text>
                </ScrollView>
                : null
              }
            </View>
          </View>
        );
    },
    wholeArea: function(){
      this.setState({
        wholeArea: true,
        hotBusiness: false,
        hotDistrict: false,
        wholeAreaFFF:{backgroundColor:'#fff'},
        hotBusinessFFF:{},
        hotDistrictFFF:{}
      });
    },
```

```
    hotBusiness: function(){
      this.setState({
        wholeArea: false,
        hotBusiness: true,
        hotDistrict: false,
        wholeAreaFFF:{},
        hotBusinessFFF:{backgroundColor:'#fff'},
        hotDistrictFFF:{}
      });
    },
    hotDistrict: function(){
      this.setState({
        wholeArea: false,
        hotBusiness: false,
        hotDistrict: true,
        wholeAreaFFF:{},
        hotBusinessFFF:{},
        hotDistrictFFF:{backgroundColor:'#fff'}
      });
    }
});
var styles = StyleSheet.create({
  container:{
    height:240,
    flex:1,
    borderTopWidth:1,
    borderBottomWidth:1,
    borderColor:'#ddd'
  },
  row:{
    flexDirection: 'row'
  },
  flex_1:{
    flex:1
  },
  header:{
    height:35,
    borderBottomWidth:1,
    borderColor:'#DFDFDF',
    backgroundColor:'#F5F5F5'
  },
  header_text:{
    color:'#7B7B7B',
    fontSize:15
  },
  center:{
    justifyContent:'center',
    alignItems:'center'
  },
  left_pannel:{
    backgroundColor:'#F2F2F2',
  },
```

```
        left_row:{
          height:30,
          lineHeight:20,
          fontSize:14,
          color:'#7C7C7C',
        },
        right_pannel:{
          marginLeft:10
        },
        active_blue:{
          color: '#00B7EB'
        },
        active_fff:{
          backgroundColor:'#fff'
        }
      });
      var App = React.createClass({
        render: function(){
          return (
            <View style={{marginTop:25}}>
              <MenuList/>
            </View>
          );
        }
      });
      AppRegistry.registerComponent('APP', () => App);
```

在上述代码中,我们首先需要规划单击一级菜单的state。这里我们使用的是一个比较简单的例子,数据是图6-1所示的数据。wholeArea表示"全部区域",hotBusiness表示"热门商圈",hotDistrict表示"热门行政区",如果它们为true,则显示对应的视图(即二级菜单)。这里,我们在点击时会控制只显示一个二级视图:

```
      getInitialState: function(){
        return {
          wholeArea: false,
          hotBusiness: true,
          hotDistrict: false,
          wholeAreaFFF:{},
          hotBusinessFFF:{backgroundColor:'#fff'},
          hotDistrictFFF:{}
        };
      },
```

其中wholeAreaFFF、hotBusinessFFF和hotDistrictFFF表示对应的"全部区域"、"热门商圈"、"热门行政区"是否为白色。如果是白色,则为{backgroundColor:'#fff'},如果不显示背景颜色,则为{}。

运行上面的代码,发现基本的静态效果已经完成。但是,如果这样就算完成的话,组件的数据模型和视图就没有分离,组件也无法重用。我们到这里充其量只是做了一个"页面"而已,而不是完成一个组件。因此，我们需要将静态组件改为动态组件。开发者可以通过传入不同的数据模型来渲染组件。

6.1.2 实现组件的复用和封装

开发一个登录系统，需要提前规划很多问题，比如：第三方url登录跳转、cookie和session的维护、统一鉴权和认证等。同样，开发一个组件时，也是需要规划的。在6.1.1节中，我们实现了二级菜单组件的静态视图，这一节需要修改代码，从而实现组件的复用。我们从以下4个方面来规划二级菜单组件。

1. 数据模型

既然要实现组件的复用，就需要将每一项的数据传递到组件中。这时数据模型（model）就十分重要，一个好的模型就是一个好的数据结构，从某种意义上说是一个好的算法。只有数据模型设计合理了，后面的模型在组件内的维护才更方便。二级菜单的模型是这样的：点击某一个Tab选项卡，展示该Tab视图的一级菜单项，点击某一级菜单项即可展示二级菜单。通过上面的描述，我们大概知道了该组件需要的树形结构模型，如图6-2所示。

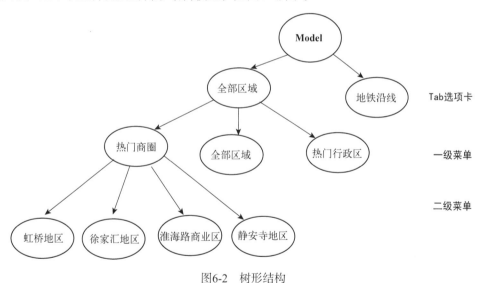

图6-2　树形结构

下面我们构建一个简单的数据模型，代码如下所示：

```
var data = {
  "全部区域": {
    "全部区域": ["全部区域"],
    "热门商圈": [
      "虹桥地区",
      "徐家汇地区",
      "淮海路商业区",
```

```
        "静安寺地区",
        "上海火车站地区",
        "浦东陆家嘴金融贸易区",
        "四川北路商业区",
        "人民广场地区",
        "南翔、安亭汽车城"
    ],
    "热门行政区": [
        "静安区",
        "徐汇区",
        "长宁区",
        "黄浦区",
        "虹口区",
        "宝山区",
        "闸北区"
    ]
},
"地铁沿线":{
    "地铁全线": ["地铁全线"],
    "一号线": ["莘庄站", "外环路站", "莲花路站", "锦江乐园站", "上海南站", "漕宝路站"],
    "二号线": ["浦东国际机场站", "海天三路站", "远东大道站", "凌空路站"]
}
};
```

此时我们已经构建好设定的数据模型,下一步就需要设计组件的属性接口了。

2. 属性接口

我们希望组件可以通过传入参数渲染不同的视图,这里希望可以按照如下接口形式使用该组件:

```
var App = React.createClass({
  render: function(){
    return (
      <View style={{marginTop:25}}>
        <MenuList data={data} nSelected={1} tabSelected={0} click={this.onPress}/>
      </View>
    );
  },
  onPress: function(val){
    alert(val);
  }
});
```

MenuList是二级菜单组件,data属性用于传递数据模型,tabSelected表示哪一个选项卡(Tab)会被选中,nSelected表示选中菜单中的哪一个子项被选中。click表示二级菜单项被点击时触发的函数。目前,我们希望该组件的接口越简单越好。因此,只暴露这4个属性。

3. 设计渲染规则

在React Native中,通过改变state来触发视图的刷新。因此,我们需要设计刷新规则,这里我们希望将model的一个子树作为视图来刷新。具体的state设计如下:

```
//设定内置的属性
//选中项，例如：_type_0_2表示第一个Tab选中并且第二个Tab中的第三项选中
var prefixType = '_type_';
//选中项样式，例如：_style_0_2表示第一个Tab选中并且第二个Tab中的第三项选中时的样式
var prefixStyle = '_style_';
//默认情况下左侧选中的背景颜色
var defaultBackgroundColor = {backgroundColor:'#fff'};
//getInitialState
getInitialState: function(){
  var data = this.props.data;
  //左侧选择的index
  var nSelected = this.props.nSelected;
  //头部选择的index
  var tabSelected = this.props.tabSelected;
  var obj = {};
  var kIndex = 0;
  for(var k in data){
    var childData = data[k];
    var cIndex = 0;
    for(var c in childData){
      var type = prefixType + k + '_' + c;
      var style = prefixStyle + k + '_' + c;
      obj[type] = false;
      obj[style] = {};
      //设定默认选中项
      if(nSelected === cIndex && tabSelected === kIndex){
        obj[type] = true;
        obj[style] = defaultBackgroundColor;
      }
      cIndex++;
    }
    kIndex++;
  }
  obj.tabSelected = tabSelected;
  obj.nSelected = nSelected;
  return obj;
}
```

在上面的代码中，以_type_为前缀的状态表示一级菜单的某一项。例如_type_0_1表示第一个Tab选项卡对应的视图一级菜单的第2项。如果其值为true，则说明被选中，否则没有选中。以_style_表示该项的样式，如果被选中，则其背景颜色为白色。我们同时将tabSelected和nSelected作为状态更新，修改_type_tabSelected_nSelected和_style_tabSelected_nSelected。这样的话，我们就可以根据点击事件刷新视图。

4. 分解渲染

分解该组件，将它分成3部分：头部Tab切换栏、左侧一级菜单、右侧二级菜单。我们希望render里面按照这3部分渲染，具体代码如下：

```
render: function(){
  var header = this.renderHeader();
  var left = this.renderLeft();
```

```
    var right = this.renderRight();
    return (
      <View style={styles.container}>
        <View style={[styles.row, styles.header]}>
          {header}
        </View>
        <View style={[styles.row, styles.flex_1]}>
          <ScrollView style={[styles.flex_1, styles.left_pannel]}>
            {left}
          </ScrollView>
          <ScrollView style={[styles.flex_1, styles.right_pannel]}>
            {right}
          </ScrollView>
        </View>
      </View>
    );
},
```

这里我们使用了renderHeader、renderLeft和renderRight这3个函数来分解渲染视图。这样做的话，我们对视图的控制能力更强。

renderHeader方法的实现如下：

```
//渲染头部TabBar
renderHeader: function(){
    var data = this.props.data;
    var tabSelected = this.state.tabSelected;
    var header = [];
    var tabIndex = 0;
    for(var i in data){
      var tabStyle = null;
      if(tabIndex === tabSelected){
        tabStyle=[styles.header_text, styles.active_blue];
      }else{
        tabStyle = [styles.header_text];
      }
      header.push(
        <TouchableOpacity style={[styles.flex_1, styles.center]}
          onPress={this.headerPress.bind(this, i)}>
          <Text style={tabStyle}>{i}</Text>
        </TouchableOpacity>
      );
      tabIndex ++;
    }
    return header;
},
```

renderLeft的实现如下：

```
//渲染左侧
renderLeft: function(){
    var data = this.props.data;
    var tabSelected = this.state.tabSelected;
    var leftPannel = [];
```

```
      var index = 0;
      for(var i in data){
        if(index === tabSelected){
          for(var k in data[i]){
            var style = this.state[prefixStyle + i + '_' + k];
            leftPannel.push(
              <Text onPress={this.leftPress.bind(this, i, k)}
                    style={[styles.left_row, style]}>  {k}</Text>);
          }
          break;
        }
        index ++;
      }
      return leftPannel;
    },
```

renderRight的实现如下：

```
//渲染右边，二级菜单
renderRight: function(){
  var data = this.props.data;
  var tabSelected = this.state.tabSelected;
  var nSelected = this.state.nSelected;
  var index = 0;
  var rightPannel = [];
  for(var i in data){
    if(tabSelected === index ){
      for(var k in data[i]){
        if(this.state[prefixType + i + '_' + k]){
          for(var j in data[i][k]){
            rightPannel.push(
              <Text onPress={this.props.click.bind(this, data[i][k][j])}
                    style={styles.left_row}>{data[i][k][j]}</Text>);
          }
          break;
        }
      }
    }
    index ++;
  }
  return rightPannel;
},
```

5. 绑定事件

我们需要在3个地方绑定：Tab切换的事件、一级菜单点击的事件、二级菜单点击的事件。

Tab切换需要做两件事：更新渲染子视图和默认选中一级菜单的第一项。首先给出头部点击事件的代码，具体如下：

```
//头部点击事件，即Tab切换事件
headerPress: function(title){
  var data = this.props.data;
  var index = 0;
```

```
          for(var i in data){
            if(i === title){
              this.setState({
                tabSelected: index,
              });
              var obj = {};
              var n = 0;
              for(var k in data[i]){
                if(n !== 0){
                  obj[prefixType + i + '_' + k] = false;
                  obj[prefixStyle + i + '_' + k] = {};
                }else{
                  obj[prefixType + i + '_' + k] = true;
                  obj[prefixStyle + i + '_' + k] = defaultBackgroundColor;
                }
                n ++;
              }
              this.setState(obj);
            }
            index ++;
          }
        }
```

然后完成一级菜单的点击事件，具体代码如下：

```
//点击左侧，展示右侧二级菜单
leftPress: function(tabIndex, nIndex){
  var obj = {};
  for(var k in this.state){
    //将prefixType或者prefixStyle类型全部置为false
    if(k.indexOf(prefixType) > -1){
      var obj = {};
      obj[k] = false;
      this.setState(obj);
    }
    if(k.indexOf(prefixStyle) > -1){
      var obj = {};
      obj[k] = {};
      this.setState(obj);
    }
  }
  obj[prefixType + tabIndex + '_' + nIndex] = true;
  obj[prefixStyle + tabIndex + '_' + nIndex] = defaultBackgroundColor;
  this.setState(obj);
},
```

二级菜单的点击事件由开发者绑定，我们只需要将被点击的数据回传给开发者即可：

```
<Text onPress={this.props.click.bind(this, data[i][k][j])}
  style={styles.left_row}>{data[i][k][j]}</Text>
```

6. 完整代码

至此，MenuList（二级菜单）组件已经开发完成。为了节省篇幅，完整代码可以参考

https://github.com/vczero/react-native-tab-menu/blob/master/tab.js，整个项目可以参考 https://github.com/vczero/react-native-tab-menu。若要使用该模块，可以使用 npm install react-native-tab 命令安装。

6.1.3 应用二级菜单组件

组件已经开发好了，现在可以使用二级菜单组件了。这里假设数据模型是编程语言（Language）和开发工具（Tool），具体的代码如下：

```
var React = require('react-native');
var MenuList = require('./MenuList');
var {
  AppRegistry,
  StyleSheet,
  Text,
  View,
  ScrollView,
  TouchableOpacity,
} = React;

var data = {
  "Language": {
    "All": ["All"],
    "Web Front End": [
      "HTML",
      "CSS",
      "JavaScript"
    ],
    "Server": [
      "Node.js",
      "PHP",
      "Python",
      "Ruby"
    ]
  },
  "Tool":{
    "All": ["All"],
    "Apple": ["Xcode"],
    "Other": ["Sublime Text", "WebStrom",]
  }
};

var App = React.createClass({
  render: function(){
    return (
      <View style={{marginTop:25}}>
        <MenuList data={data} nSelected={1} tabSelected={0}
          click={this.onPress}/>
      </View>
    );
  },
```

```
    onPress: function(val){
      alert(val);
    }
  });
```

AppRegistry.registerComponent('APP', () => App);

运行该组件的效果如图6-3所示。

图6-3　二级菜单组件

6.2　日历组件

React Native给我们提供了很多原生的组件，它们的性能和体验都很好。但是，有些组件不太符合产品的需求，我们需要自己重新定义组件，比如经常用到的日历组件。一般情况下，日历组件都是"全页面"日历，如图6-4所示。

图6-4　全页面日历

6.2.1 开发日历组件

日历组件的开发重点是计算日历的逻辑。我们需要知道日历显示多少行,每月的第一天从星期几开始。一个星期7天,一个月最多31天,这是基本常识。但是,我们会误以为7(天)×5(天)就足以表达一个月份的日历。如图6-5所示,左图是5行的日历主体,右图是6行的日历主体,这是我们开发时需要注意的。

图6-5 日历主体

现在我们对日历有了个初步的认识,接下来就开始开发日历组件。

1. 确定需求

开发组件前,我们需要想好日历的组件暴露的接口和实现的功能。这里,我们需要完成的日历的接口和功能如下所示。

- 全页面日历的实现,可以传入参数,显示多少个月份,默认3个。
- 可以定义日历开始的日期,比如2015-7-8。尽管该日期已过,但是日历会从7月开始显示。默认的开始日期是今天。
- 历史日期以灰色字体显示,从今天开始的日期以黑色字体显示。
- 可以传入选中日期,该选中日期会高亮显示(背景色为蓝色,字体为白色)。
- 可以定义星期栏的字体大小和背景颜色。
- 可以显示节假日。
- 点击日期获取日期字符串。

目前,按照以上需求,该日历组件应该可以满足基本的业务需要。

2. 确定接口

根据以上需求,日历组件需要提供一个接口供开发者传入参数。我们希望开发人员这样使用日历组件:

```
render: function(){
  var holiday = {
    '2015-10-1': '国庆节',
    '2015-9-10': '教师节',
    '2016-1-1': '元旦节',
```

```
    '2015-11-11': '双十一'
};
var check = {
    '2015-10-1': 'checked',
    '2015-9-1': 'checked',
    '2015-7-10': 'checked',
    '2015-9-10': 'checked'
};
var headerStyle ={
    backgroundColor: '#3C9BFD',
    color:'#fff',
    fontSize: 15,
    fontWeight:500,
};
return (
    <View style={styles.container}>
        <Calendar
          touchEvent={this.press}
          headerStyle={headerStyle}
          holiday={holiday}
          startTime={new Date(2015, 6, 8)}
          check={check}
          num={5}
        />
    </View>
);
}
```

在上述代码中，num={5}表示全页面日历显示5个月份。holiday表示显示节假日，是JavaScript对象，key值是日期，value为显示的值。headerStyle用于设置头部的样式。startTime表示日历显示的开始时间。check表示选中的日期，一般会高亮显示。

3. 初始化数据模型

接口可以给我们传递参数，但是我们需要对传递的参数进行模型转换，从而变成需要的model对象。这在getInitialState中处理：

```
getInitialState: function(){
    //开始时间
    var startTime = this.props.startTime || new Date();
    var holiday = this.props.holiday || {};
    var check = this.props.check || {};
    var headerStyle = this.props.headerStyle || {};
    //显示月份的个数
    var num = this.props.num || 3;
    return {
        startTime: startTime,
        num: num,
        holiday: holiday,
        check: check,
        headerStyle: headerStyle
    };
},
```

4. 计算日历的行数

在开篇已经知道，日历的行数十分重要，因为这关系到渲染日历主体需要循环几次。那么，日历的行数如何计算呢？具体如下：

```
日历行数 = （空白格数 + 月份日期占用的格数即天数） / 7
```

那么，空白格数如何计算呢？我们不关注月末日期后面的空格，只关注1号前面的空格数。那么，如何才能知道前面有几个空格呢？这个比较好办，就是1号是星期几，就说明前面存在（星期几-1）个空格：

```
空格数 = 1号星期几 - 1;
```

最后的目标是希望算出该月份有多少行日期。综合上面的两个等式，可以得出：

```
日历行数 = Math.ceil((月份的天数 + 1号星期几 - 1) / 7)
```

之所以向上取整，主要有两个原因：行数是整数；当出现小数点，说明必须多开闭一个空格给日期占用，也就是多开闭一行。转化成代码就是：

```
var date = this.state.startTime;
var num = this.state.num;
var holiday = this.state.holiday;
var check = this.state.check;
var headerStyle = this.state.headerStyle;
var items = [];
var dateNow = new Date();
for(var n = 0; n < num; n++){
  /*循环完成一个月*/
  var rows = [];
  var newDate = new Date(date.getFullYear(), date.getMonth() + 1 + n, 0); //天数
  var week = new Date(date.getFullYear(), date.getMonth() + n, 1).getDay(); //月份开始的星期
  if(week === 0){
    week = 7;
  }
  var counts = newDate.getDate();
  var rowCounts = Math.ceil((counts + week - 1) / 7); //本月行数
  //TODO: 渲染日期行
}
```

5. 渲染视图

我们已经知道了日历行数，现在渲染日历视图就要容易得多。这里的日历渲染实际上是3个for循环。第一层循环是全页面展示几个月份，第二层循环是每个月渲染几行，第三层是渲染每个日期。这里之所以单独渲染每个日期，是因为有时候需要给日期加上标识，比如节假日的显示。主体的逻辑代码如下：

```
render: function() {
var date = this.state.startTime;
var num = this.state.num;
var holiday = this.state.holiday;
```

```
var check = this.state.check;
var headerStyle = this.state.headerStyle;
var items = [];
var dateNow = new Date();
for(var n = 0; n < num; n++){
  /*循环完成一个月*/
  var rows = [];
  var newDate = new Date(date.getFullYear(), date.getMonth() + 1 + n, 0); //天数
  var week = new Date(date.getFullYear(), date.getMonth() + n, 1).getDay(); //月份开始的星期
  if(week === 0){
    week = 7;
  }
  var counts = newDate.getDate();
  var rowCounts = Math.ceil((counts + week - 1) / 7); //本月行数
  for(var i = 0; i < rowCounts; i++){
    var days = [];
    for(var j = (i * 7) + 1; j < ((i+1) * 7) + 1; j++){
      //根据每个月开始的 [星期] 往后推
      var dayNum = j - week + 1;
      if(dayNum > 0 && j < counts + week){
        //如果当前日期小于今天，则变灰
        var dateObj = new Date(date.getFullYear(),
          date.getMonth() + n, dayNum);
        var dateStr = dateObj.getFullYear() + '-' +
          (dateObj.getMonth() + 1) + '-' + dayNum;
        var grayStyle = {};
        var bk = {};
        if(dateNow >= new Date(date.getFullYear(),
          date.getMonth() + n, dayNum + 1)){
          grayStyle = {
            color:'#ccc'
          };
        }
        if(holiday[dateStr]){
          dayNum = holiday[dateStr];
        }
        if(check[dateStr]){
          bk = {
            backgroundColor: '#1EB7FF',
            width:46,
            height:35,
            alignItems: 'center',
            justifyContent: 'center'
          };
          grayStyle = {
            color:'#fff'
          };
        }
        days.push(
          <TouchableHighlight style={[styles.flex_1]}
            underlayColor="#fff" onPress={this.props.touchEvent?
            this.props.touchEvent.bind(this, dateStr):null}>
            <View style={bk}>
              <Text style={grayStyle}>{dayNum}</Text>
```

```
          </View>
        </TouchableHighlight>
      );
    }else{
      days.push(
        <View style={[styles.flex_1]}>
          <Text></Text>
        </View>
      );
    }
  }
  rows.push(
    <View style={styles.row}>{days}</View>
  );
}
items.push(
  <View style={[styles.cm_bottom]}>
    <View style={styles.month}>
      <Text style={styles.month_text}>{newDate.getFullYear()}年
      {newDate.getMonth() + 1}月</Text>
    </View>
    {rows}
  </View>
);
}
return (
  <View style={styles.calendar_container}>
    <View style={[styles.row, styles.row_header, this.props.headerStyle]}>
      <View style={[styles.flex_1]}>
        <Text style={this.props.headerStyle}>一</Text>
      </View>
      <View style={[styles.flex_1]}>
        <Text style={this.props.headerStyle}>二</Text>
      </View>
      <View style={[styles.flex_1]}>
        <Text style={this.props.headerStyle}>三</Text>
      </View>
      <View style={[styles.flex_1]}>
        <Text style={this.props.headerStyle}>四</Text>
      </View>
      <View style={[styles.flex_1]}>
        <Text style={this.props.headerStyle}>五</Text>
      </View>
      <View style={[styles.flex_1]}>
        <Text style={[styles.week_highlight, this.props.headerStyle]}>六</Text>
      </View>
      <View style={[styles.flex_1]}>
        <Text style={[styles.week_highlight, this.props.headerStyle]}>日</Text>
      </View>
    </View>
    <ScrollView style={{flex:1,}}>
      {items}
    </ScrollView>
  </View>
```

```
    );
}
```

6. 完整的组件代码

经过上面的分析，我们大致完成了日历组件。最后，可以使用module.exports将组件暴露出去。目前，组件已经发布到npm上。如果以后要使用该日历组件，可以直接使用npm installrn-calendar命令。为了节省篇幅，不再附完整代码。若需查看完整的代码，可以参考 https://github.com/vczero/react-native-calendar/blob/master/calendar.js。

6.2.2 应用日历组件

现在已经开发好日历组件，接下来就可以使用它了。我们显示从今天开始的10个月份，并且显示自定义的节假日和高亮显示一些日期。目前，日历组件已经发布在npm上了，名称为rn-calendar。所以直接使用npm install rn-calendar命令安装即可。演示代码如下所示：

```
var React = require('react-native');
var Calendar = require('rn-calendar');
var{
  View,
  AppRegistry,
  StyleSheet,
  StatusBarIOS
} = React;

StatusBarIOS.setHidden(true);

var Index = React.createClass({
  render: function(){
    var holiday = {
      '2015-10-1': '国庆节',
      '2015-9-10': '教师节',
      '2016-1-1': '元旦节',
      '2015-11-11': '双十一'
    };
    var check = {
      '2015-10-1': 'checked',
      '2015-9-1': 'checked',
      '2015-7-10': 'checked',
      '2015-9-10': 'checked',
      '2015-11-11': 'checked'
    };
    var headerStyle ={
      backgroundColor: '#3C9BFD',
      color:'#fff',
      fontSize: 15,
      fontWeight:500,
    };
    return (
      <View style={styles.container}>
        <Calendar
```

```
                touchEvent={this.press}
                headerStyle={headerStyle}
                holiday={holiday}
                startTime={new Date()}
                num={10}
                check={check}
                />
        </View>
    );
  },
  press: function(str){
    alert(str);
  }
});
var styles = StyleSheet.create({
  container: {
    flex: 1,
    backgroundColor:'blue'
  }
});
AppRegistry.registerComponent('APP', () => Index);
```

上述代码的运行效果如图6-6所示。

图6-6 日历组件运行效果

6.3 开源组件

这里我们已经造了两个轮子：二级菜单组件和日历组件。有些时候，因为项目紧张，没有时间去造轮子,这时候需要寻找开源组件。一般情况下,我们都习惯在GitHub中搜索开源仓库。在使用开源组件的时候,一定要先测试,确保组件稳定可用。如果开源组件不满足需求,需要有能力修改组件源码以满足自己的需求。在这一节中,我们介绍两个常用的开源组件。

6.3.1 react-native-swiper

一款APP,尤其是内容主导的APP都存在一个组件——轮播组件。我们在GitHub上找到了react-native-swiper组件。细致地看下react-native-swiper的文档,发现该组件满足我们做轮播组件的需求。现在我们使用react-native-swiper来开发一个示例。

1. 安装react-native-swiper

我们可以通过npm安装react-native-swiper。在项目的根目录(即package.json文件所在的目录)开启终端,输入如下命令:

```
$ npm i react-native-swiper --save
```

等待一下,会发现安装成功。

2. 演示示例

安装成功后,需要确保组件能够正常使用。我们从GitHub上复制下来,具体代码如下:

```
var Swiper = require('react-native-swiper');
var React = require('react-native');
var {
  AppRegistry,
  StyleSheet,
  Text,
  View,
} = React;

var swiper = React.createClass({
  render: function() {
    return (
      <View>
        <Swiper style={styles.wrapper} showsButtons={true}>
          <View style={styles.slide1}>
            <Text style={styles.text}>第一张</Text>
          </View>
          <View style={styles.slide2}>
            <Text style={styles.text}>第二张</Text>
          </View>
          <View style={styles.slide3}>
            <Text style={styles.text}>第三张</Text>
          </View>
        </Swiper>
      </View>
    );
  }
});
```

```
var styles = StyleSheet.create({
  wrapper: {
  },
  slide1: {
    flex: 1,
    justifyContent: 'center',
    alignItems: 'center',
    backgroundColor: '#9DD6EB',
  },
  slide2: {
    flex: 1,
    justifyContent: 'center',
    alignItems: 'center',
    backgroundColor: '#97CAE5',
  },
  slide3: {
    flex: 1,
    justifyContent: 'center',
    alignItems: 'center',
    backgroundColor: '#92BBD9',
  },
  text: {
    color: '#fff',
    fontSize: 30,
    fontWeight: 'bold',
  }
});

AppRegistry.registerComponent('APP', () => swiper);
```

我们在Xcode上使用快捷键cmd+R运行项目，编译并启动成功。注意：如果使用npm install 安装了新的组件，需要在Xcode上重新编译启动，以加载node_module里面全部的库。如果只是在模拟器上使用快捷键cmd+R刷新，是不会加载刚安装的组件的。示例的运行效果如图6-7所示。

图6-7　轮播

3. 开发图片轮播功能

刚才已经看到轮播的效果，现在需要使用react-native-swiper来开发图片轮播组件。其实该组件已经包含了很多功能和API。我们使用既有的API即可。我们的代码如下：

```
var swiper = React.createClass({
  render: function() {
    return (
      <View style={{marginTop:23}}>
        <Swiper style={styles.wrapper} showsButtons={false} autoplay={true} height={200}>
          <View style={styles.slide1}>
            <Image
              style={{height:200,width:320}}
              resizeMode="stretch"
              source={{uri:'http://vczero.github.io/ctrip/lvtu/img/2.jpg'}}/>
          </View>
          <View style={styles.slide2}>
            <Image
              style={{height:200,width:320}}
              resizeMode="stretch"
              source={{uri:'http://vczero.github.io/ctrip/lvtu/img/city.jpg'}}/>
          </View>
          <View style={styles.slide3}>
            <Image
              style={{height:200, width:320}}
              resizeMode="stretch"
              source={{uri:'http://vczero.github.io/ctrip/lvtu/img/4.jpg'}}/>
          </View>
        </Swiper>
      </View>
    );
  }
});
```

这里我们没有将样式代码独立出来是为了方便贴代码，节省空间。在Swiper组件中主要使用了如下属性。

- **showsButtons={false}**：隐藏左右滑动的箭头。
- **autoplay={true}**：自动轮播。
- **height={200}**：设置轮播组件的包裹容器高度。注意：这里需要使用属性设置，不能使用样式设置。

同时，我们将Text组件替换成了组件Image，循环播放3张图片。其实图片的大小可以使用Dimensions获取屏幕的大小来设置。这里，我们简单地将其设置为320。因为使用了开源组件，所以图片轮播的功能可以轻松完成。运行效果如图6-8所示。关于react-native-swiper的更多API，可以参考https://github.com/leecade/react-native-swiper。

图6-8　图片轮播效果

4. 源码阅读

使用react-native-swiper开发图片轮播功能，十分方便。但是，我们希望了解如何实现滚动视图？是动画隐藏视图还是使用了ScrollView？我们在GitHub上找到react-native-swiper的源码，发现源码并不是很多，但是设计还是很巧妙，如下所示：

```
return (
  <View style={[styles.container, {
    width: state.width,
    height: state.height
  }]}>
    <ScrollView ref="scrollView"
      {...props}
      contentContainerStyle={[styles.wrapper, props.style]}
      contentOffset={state.offset}
      onScrollBeginDrag={this.onScrollBegin}
      onMomentumScrollEnd={this.onScrollEnd}>
      {pages}
    </ScrollView>
    {props.showsPagination && (props.renderPagination
      ? this.props.renderPagination(state.index, state.total, this)
      : this.renderPagination())}
    {this.renderTitle()}
    {this.props.showsButtons && this.renderButtons()}
  </View>
)
```

可以发现，react-native-swiper组件使用ScrollView作为轮播组件的基础，通过计算步长滚动。更多细节，可前往https://github.com/leecade/react-native-swiper/blob/master/src/index.js了解。

6.3.2　react-native-modal

我们在使用支付宝支付的时候，会看到有一个支付的遮罩层，这样用户就可以只关心支付，而不会操作支付宝的其他功能。在React Native v0.10之前，是不存在模态对话框的。GitHub上有一个开源的组件——react-native-modal，它的API比较丰富，使用react-native-modal开发模态对话框相关的功能是一个不错的选择。

这里我们来简单实现火车票购票的支付模态对话框。之所以选择react-native-modal，主要有两个原因：一是该组件的API相对丰富，二是可以很好地演示需要编译的开源组件如何使用。我们按照如下步骤来实现该功能。

1. 安装react-native-modal

通过阅读GitHub上的文档和react-native-modal的源码，我们知道该组件依赖的第三方组件是包含Objective-C代码的，因此需要编译。我们首先使用npm安装该组件。

在项目的根目录下（即package.json和node_modules同级目录）打开终端，输入如下命令：

```
npm install react-native-modal --save
```

安装完成后，我们需要添加该组件到项目中，具体步骤如下。

（1）打开Xcode，右击项目的Libraries，选择Add Files to 'Your Project Name'，然后选择node_modules/react-native-modal/RNModal.xcodeproj。

（2）选择项目，在右边选择Build Phases → Link Binary With Libraries，选择libRNModal.a即可。

2. 使用react-native-modal

react-native-modal的API有很多，这里我们使用isVisible和onClose事件监听。我们将需要的视图放在Modal里即可，完整的代码如下：

```
var React = require('react-native');
var Modal = require('react-native-modal');
var {
  AppRegistry,
  StyleSheet,
  View,
  Navigator,
  TouchableOpacity,
  Text,
  PixelRatio
} = React;

var CloseBtn = React.createClass({
  render: function(){
    return(
      <View>
        <Text>关闭</Text>
      </View>
    );
  }
});

var App = React.createClass({
  getInitialState: function(){
    return{
      isModalOpen: true
    };
```

```
    },

    openModal: function(){
      this.setState({isModalOpen: true});
    },

    closeModal: function(){
      this.setState({isModalOpen: false});
    },

    render: function(){
      return (
        <View style={styles.page}>
          <Text onPress={() => this.openModal()}>
              预定火车票
          </Text>
          <Modal
            isVisible={this.state.isModalOpen}
            onClose={() => this.closeModal()} >
              <View style={styles.zhifu}>
                <Text style={styles.date}>2015/10/01</Text>
                <View style={styles.row}>
                  <View style={styles.point}>
                    <Text style={styles.station}>上海站</Text>
                    <Text style={styles.mp10}>8:00</Text>
                  </View>
                  <View>
                    <Text style={styles.at}></Text>
                    <Text style={[styles.mp10,{textAlign:'center'}]}>
                      G321
                    </Text>
                  </View>
                  <View style={styles.point}>
                    <Text style={[styles.station, {textAlign:'right'}]}>北京站</Text>
                    <Text style={[{textAlign:'right'}, styles.mp10]}>12:35</Text>
                  </View>
                </View>

                <View style={styles.mp10}>
                  <Text>票价：¥500.0元</Text>
                  <Text>乘车人：王**</Text>
                  <Text>上海站 2层火车厅 15检票口</Text>
                </View>
                <View style={[styles.mp10,{alignItems:'center'}]}>
                  <View style={styles.btn}>
                    <Text style={styles.btn_text}>去支付</Text>
                  </View>
                </View>
              </View>
          </Modal>
        </View>
      );
    }
});
```

```
var styles = StyleSheet.create({
  page: {
    flex: 1,
    position: 'absolute',
    bottom: 0,
    left: 0,
    right: 0,
    top: 23
  },
  zhifu:{
    height: 150,
  },
  row:{
    flexDirection:'row'
  },
  point:{
    flex:1,
    fontSize:20,
  },
  at:{
    color:'#3BC1FF',
    borderWidth:1 / PixelRatio.get() ,
    width:80,
    borderColor:'#18B7FF',
    height:1,
    marginTop:10
  },
  date:{
    textAlign:'center',
    marginBottom:5
  },
  station:{
    fontSize:20
  },
  mp10:{
    marginTop:5
  },
  btn:{
    width:60,
    height:30,
    borderRadius:3,
    backgroundColor:'#FFBA27',
    padding:5,
  },
  btn_text:{
    lineHeight:18,
    textAlign:'center',
    color:'#fff',
  }
});

AppRegistry.registerComponent('APP', () => App);
```

整个代码的运行效果如图6-9所示。

图6-9 火车票预订

第三部分
App 更新和上架篇

- 第7章　热更新和上架

第 7 章 热更新和上架

在iOS应用开发中，令我们最难以忍受的就是每次发布新版本时漫长的审核期。对于大多数敏捷开发团队来说，为了能够快速出产品、出成果，一个版本的迭代周期往往缩短在几天的时间里。所以经常会遇到当前版本还未审核通过，新版本就已经开发完成的尴尬局面，因此开发团队不得不将版本发布周期改为两周甚至一个月，以迎合苹果效率低下的评审周期。而React Native中的JavaScript代码是按动态加载执行的，因此从服务器端动态更新JavaScript代码来实现应用的更新成为了可能。当然，如果是涉及React Native框架本身的更新，还要通过重新发布应用来完成。本章中，我们会介绍一种动态更新的方法。此外，对于刚上手的同学，本章也会介绍苹果应用上架的流程。

7.1 动态更新

在传统Web开发中我们在JavaScript中修改完逻辑，直接在浏览器上刷新就可以看到效果了。之所以能这么做，首先因为JavaScript是一门动态语言，并不需要编译，然后浏览器每次刷新都动态地从服务端加载JavaScript文件。相信大家都知道，针对Web应用优化中最简单也是最有效的就是缓存，当服务端没有更新，那浏览器请求时则无须每次重新下载整个文件。我们在React Native实现的动态更新也是同样的思路，React Native中JavaScript代码最终都会打包成一个jsbundle文件，我们只在需要更新的时候，在应用中从远程服务器上下载这个文件，并重新加载，就可以完成动态更新同时无须通过App Store重新发布。试想一下，在一个电商应用中可以随时修改我们的商品界面，例如设计布局，并且绕过漫长的审核周期，这样的改变无疑是非常有价值的。

7.1.1 初始化设置

在React Native的生成代码中，默认会提示我们将jsCodeLocation设置为远程服务地址或者本地资源的路径：

```
//直接引用远程文件
```

```
jsCodeLocation = [NSURL URLWithString:@"http://localhost:8081/index.ios.bundle"];

//从本地资源读取
jsCodeLocation = [[NSBundle mainBundle] URLForResource:@"main" withExtension:@"jsbundle"];
```

如果要实现动态更新，最简单的方法就是以采用远程文件的方式来提供jsCodeLocation。但是，这样做也有一些弊端，例如首次使用的延时、网络劫持等行为也可能让应用出现意外。因此，这里建议使用维护本地文件的方式来实现热更新。首先，需要动态设置jsCodeLocation。这里我们创建一个Native API组件VersionManager来实现动态更新。每一个应用的版本都需要为本地资源的bundle文件设置一个初始版本，每次应用启动时通过[VersionManager currentVersionPath]方法来动态返回当前版本的jsCodeLocation：

```
//AppDelegate.m
- (BOOL)application:(UIApplication *)application
  didFinishLaunchingWithOptions:(NSDictionary *)launchOptions
{
...
//设置内置资源版本号
[VersionManager setInAppVersion:@"FIRST_VERSION"];
RCTBridge *bridge = [[RCTBridge alloc] initWithDelegate:self launchOptions:launchOptions];
RCTRootView *rootView = [[RCTRootView alloc]
  initWithBridge:bridge moduleName:@"..." initialProperties:nil];
...

}

//RCTBridge委托方法，返回当前jsbundle文件的地址
- (NSURL *)sourceURLForBridge:(__unused RCTBridge *)bridge
{
  return [VersionManager currentVersionPath];
}
```

每次成功更新jsbundle文件后，都需要缓存最新的版本号以及上一个版本号。首先，缓存最新版本号的目的很简单，就是为了获取当前版本的JavaScript文件的路径。因此，对于jsbundle文件的本地缓存沙盒路径，必须是可以通过版本号来定位获取的，例如这样的存储路径：

```
/Documents
    |---- ReactNative
        | --- 1.0.0
            |---- main.jsbundle
        | --- 1.1.0
            |---- main.jsbundle
```

这里我们缓存上一个版本号，这是因为如果发布了一个有缺陷的版本，可能会导致应用崩溃，此时则需要回滚至上一个版本来保证用户正常使用。

最后，我们需要有一个远端Restful API服务，通过访问这个远端服务来获得当前应用版本支持最新的bundle文件的版本号。版本服务器端大致提供了如下几个信息：最新版本、版本地址和

校验码。相关代码如下：

```
{
  ..
  "latestVersion":"2.0",
  "path":"...",
  "checksum":"..."
  ...
}
```

7.1.2　更新逻辑

更新的大致流程如下。

（1）通过Restful服务获得当前App版本支持的最新的bundle文件版本。

（2）将模块中缓存的当前版本信息与之对比，如果无须更新，则中止流程。

（3）如果需要更新，则通过之前Restful服务请求的返回数据的新版本bundle文件的地址来下载并校验更新文件，将文件存储在指定的目录下，同时处理版本信息的维护，更新当前版本号以及上一个版本号。

在我们的API组件中，需要实现以下功能：

```
//VersionManager.h

//返回当前路径
+ (NSURL *)currentVersionPath;

//访问远程版本的配置信息
- (id)accessRemoteConfig;

//下载更新
- (id)downloadBundle:(NSString *)version;
```

在更新策略上，我们可以根据产品需要来选择不同的策略。

可以采取静默的方式实现动态更新。我们可以在原生模块中监听App的运行状态，在启动时静默检测，并下载和校验更新包完成更新，此时在应用下一次启动时，将会自动使用最新版本的jsbundle文件：

```
- (instancetype)init
{
  ...
  //在应用启动时检测更新
  [[NSNotificationCenter defaultCenter] addObserver:self selector:@selector(appDidLaunch:)
    name:UIApplicationDidFinishLaunchingNotification object:nil];
  ...
}
```

```objc
- (void)appDidLaunch:(id)noti
{
    //检查远程版本配置，并更新JavaScript文件
}
```

此外，也可以提供给用户，由用户来决定是否需要更新文件并重新加载。我们需要提供一些API接口暴露给JavaScript调用：

```objc
//获得当前版本信息
RCT_EXPORT_METHOD(getCurrentVersion:(RCTPromiseResolveBlock)resolve
                      rejecter:(RCTPromiseRejectBlock)reject) {
    ...
}

//检查远程版本配置
RCT_EXPORT_METHOD(getLatestVersion:(RCTPromiseResolveBlock)resolve
                      rejecter:(RCTPromiseRejectBlock)reject) {
    ...
}

//下载JavaScript
RCT_EXPORT_METHOD(downloadVersion:(NSString *)version
                      resolver:(RCTPromiseResolveBlock)resolve
                      rejecter:(RCTPromiseRejectBlock)reject)
{
    ...
}

//重新加载JavaScript
RCT_EXPORT_METHOD(reloadVersion:(NSString *)version
                      resolver:(RCTPromiseResolveBlock)resolve
                      rejecter:(RCTPromiseRejectBlock)reject)
{
    ...
}
```

这样就可以在JavaScript中调用模块方法来完成更新流程：

```javascript
VersionManager.getLatestVersion()
.then( (config) => {
  ...
  //如果存在新版本，则下载
  VersionManager.downloadVersion(config.latestVersion)
  .then(
    //校验文件，提示更新
    ...
  ).catch(
    ...
  );
})
.catch(
```

```
    (err) => {console.log(err);
});
```

7.1.3　回滚策略

在日常的开发中，纵然有多道程序去测试，也免不了有时候发布上去的版本会存在这样那样的问题。传统的解决方案是将之前的版本重新发布一次。但是我们也知道发布流程非常耗时，尤其是大公司，流程走的更多，回滚需要的时间也会更长。这里我们不解决所有的回滚问题，主要阐述一种bundle文件中报错的回滚机制。在调试模式下，我们可以在控制台中看到错误，也好定位和分析问题出现在哪里。但在生产环境下，就比较麻烦了，会直接导致应用崩溃。这里我们主要阐述生产环境下bundle中报错的回滚机制。我们通过原生代码去监控bundle文件。当应用崩溃时，先要判断出是否是React Native跑出的异常，同时定义自己的回滚策略，例如当崩溃次数与启动次数的比值达到多少时执行回滚操作。根据前面的介绍，每次更新完的bundle文件都会有两个版本并存，即当前版本和上一个版本。如果当前版本出现问题，我们不需要重新发布代码，App中会自动将地址切换到上一个版本，以达到代码回滚。

下面是简单的代码实现：

```
//AppDelegate.m

- (BOOL)application:(UIApplication *)application
  didFinishLaunchingWithOptions:(NSDictionary *)launchOptions
{
  NSSetUncaughtExceptionHandler(&uncaughtExceptionHandler);
  ...
}

void uncaughtExceptionHandler(NSException *exception) {
  //判断JavaScript异常，处理回滚策略。
  //处理版本信息，回退至上一个版本
  [VersionManager revertToPrevious];

}
```

通过上面的学习，一个简单的热更新流程就算完成了。当然，里面还有很多需求和细节需要我们不断完善，但是大体思路应该就是这样。另外，在动态更新机制带来的优势下，代码安全性的劣势也将会被进一步放大，核心代码的安全将会是未来大家关注的一个问题。

7.2　App 上架

学习到这里的时候，我们相信你完全可以用本书学到的知识来开发一个完整的App了。但是，开发App的最终目的还是要给用户来使用才能体现出它的价值。在开发的过程中，我们一般会经历这么几个阶段。首先在模拟器中调试，然后再使用真机调试，发布测试，最后才是正式发布在App Store上。在这一节中，我们就来学习App开发中的最后一个环节——App的发布，即如何将App发布在App Store上。

7.2.1 证书生成

应用要成功发布，首先需要准备以下环节：

- 生成发布证书；
- 生成应用的唯一标识App ID；
- 生成发布证书、App ID对应的描述文件。

整个过程略有些复杂，下面为大家详细介绍一下。

1. 生成发布证书

首先，我们的账号必须加入苹果开发者计划。登录Member Center界面，进入Certificates, Identifiers & Profiles，如图7-1所示。

图7-1　用户中心界面

在打开的界面中选择Certificates证书菜单，如图7-2所示。

图7-2　Certificates界面

然后进入了Certificates, Identifiers & Profiles页面，在这个页面中可以生成证书，并设置App ID以及描述文件。

我们先生成证书。点击Certificates菜单下的Production证书类型，然后点击右上方的添加符号，新增一个证书，如图7-3所示。

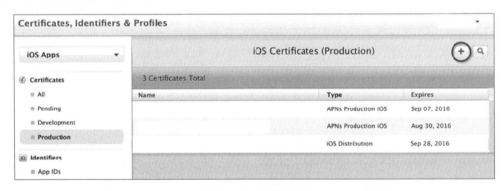

图7-3　添加证书界面

此时就会进入一个注册之类的流程界面，引导我们一步步完成证书生成。这里是为了让应用上架，所以选择App Store and Ad Hoc项，如图7-4所示。

```
Production

○ App Store and Ad Hoc
   Sign your iOS app for submission to the App Store or for Ad Hoc distribution.

○ Apple Push Notification service SSL (Production)
   Establish connectivity between your notification server and the Apple Push Notification service
   production environment. A separate certificate is required for each app you distribute.

○ Pass Type ID Certificate
   Sign and send updates to passes in Wallet.
```

图7-4　选择证书类型界面

接着点击"下一步"按钮，此时会看到一个讲述如何生成证书签名CSR文件的用户手册。看完之后继续，此时会要求我们上传CSR文件，如图7-5所示。

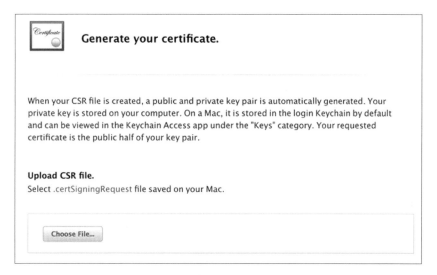

图7-5　上传证书签名文件界面

根据之前界面中介绍的流程，我们需要通过钥匙串来生成一个证书签名文件。

我们可以通过在Spotlight中键入keychain或者"应用程序"→"实用工具"→"钥匙串访问"菜单来进入钥匙串界面，然后选择"钥匙串访问"→"证书助理"→"从证书颁发机构请求证书..."菜单，如图7-6所示。

图7-6 请求证书选择界面

此时会要求填写用户电子邮件,这可以用开发者账号的注册邮箱。常用名称可以随意填写。最后选择"存储到磁盘",生成文件,如图7-7所示。

图7-7 证书信息填写界面

这样我们的证书签名文件就算生成了。接下来,继续生成应用发布证书。回到图7-5所示的界面,选择并上传生成的证书签名文件,成功后就会跳转图7-8所示的证书下载界面,根据页面上介绍的流程,下载并且点击证书文件,完成证书的安装。这里需要注意的是,如果第三方希望使用这个发布证书,直接从账户界面中下载证书将会因为缺少密钥而无法使用。因此,在安装完证书后,我们会从钥匙串中备份成功安装的证书,而备份文件中将会包含证书以及密钥。第三方只需要通过这个备份文件来安装证书就可以正常使用了。

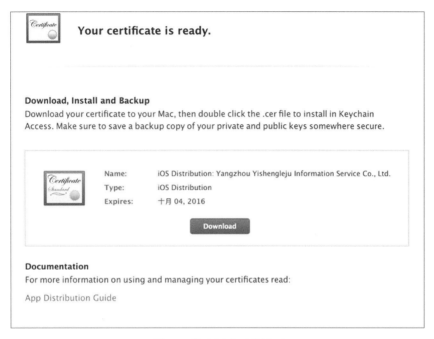

图7-8　发布证书下载界面

2. 注册App ID

App ID是苹果设备识别应用的唯一标识。回到图7-3所示的界面中，选择Identifiers→App IDs，然后点击右上方的添加按钮，进入注册页面，将会要求填入以下信息。

- App ID Description：描述，随便填。
- App ID Prefix：苹果自动生成的TeamID。
- App ID Suffix：这里就是我们为应用定义的标识，需要与应用中的Bundle Identifier一致。这里允许通过Wildcard App ID使用*来匹配多个Bundle Identifier。

在键入应用的Bundle Identifier后，苹果会自动监测该标识是否重复。完成表单的填写后，成功注册App ID。

3. 添加描述文件

接着要创建描述文件了。回到图7-3所示的界面，选择Provisioning Profiles→Distribution项，点击添加按钮进入新建描述文件的流程。这里选择Distribution→App Store，如图7-9所示。

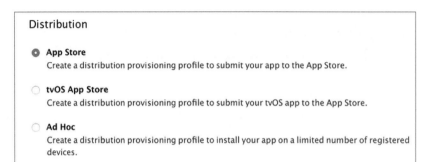

图7-9　Distribution界面

点击"继续"按钮后，会让我们选择App ID，如图7-10所示。

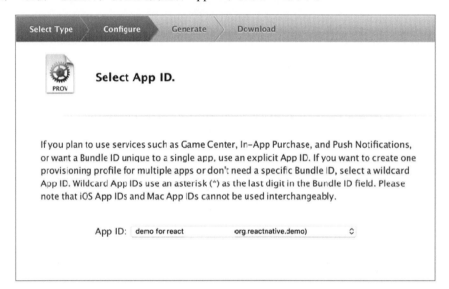

图7-10　选择App ID界面

接着会让选择发布证书，如图7-11所示。

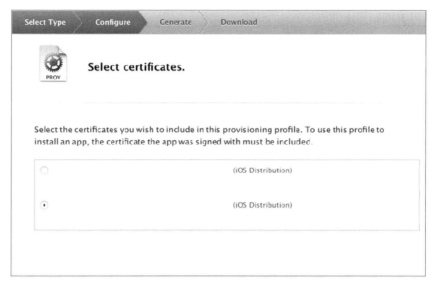

图7-11　发布证书界面

这里会让我们选择App ID以及发布证书。这两项前面已经成功添加了，在导航页面中会有单选按钮将它们显示出来。选择之前生成的App ID以及证书，即可生成描述文件。最后，下载并双击描述文件，完成安装，如图7-12所示。

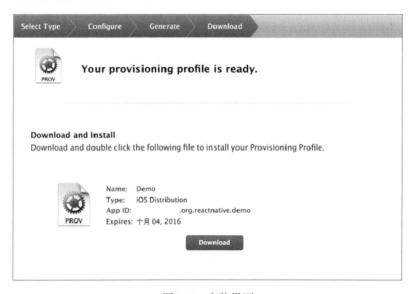

图7-12　安装界面

7.2.2 注册应用

在这一节中，我们注册需要上架的应用。回到开发者首页（即图7-1所示的用户中心界面），进入iTunes Connect来管理我们的应用发布。选择"我的App"后，点击左上角的添加符号，选择"新App"项来新增一个应用，如图7-13所示。

图7-13　添加应用界面

此时会打开应用信息填写界面，如图7-14所示。

图7-14　应用信息填写界面

先为应用填写一些基本信息，如果不了解，可以点击标题后面的问号查看。点击"套装ID"下拉按钮，可以看到前面注册的App ID，选择对应的App ID，然后点击"创建"按钮后，此时就创建好新的应用了。

此外，还要添加以下信息。

❏ **设置价格**：为对应地区设置应用下载的价格。

❑ 版本信息：包括版权信息、版本、Icon图标、分级信息、屏幕快照、描述、关键词、技术支持网址、营销网址等，这些内容均会显示在App Store的详情中。
❑ 审核信息：包括联系信息、演示账户。审核相关的问题都会使用此处填写的联系方式。
❑ 版本发布设置：苹果提供了一个作业任务，允许用户在审核通过后自动、手动以及定时上架应用。

图7-15　新应用添加完成界面

如果你的应用包含一些内部购买以及Game Center的设置，也可以在这里进行设置，这里就不详细介绍了。

苹果对于应用的审核比较严格，某些信息的缺失将无法通过审核，因此请尽量完整地完成信息的填写。

7.2.3　上传应用

设置完应用信息后，就可以通过Xcode上传应用了，具体步骤如下。

（1）选择项目工程，在General→Identify中确保Bundle Identifier与发布证书的App ID一致，如图7-16所示。

图7-16　Identifier界面

(2) 进入Build Settings→Code Signing，在Code Signing Identifyd的Release中选择对应的发布证书签名，同时在描述文件Provisioning Profile中选择之前安装的文件，如图7-17所示。

图7-17　Code Signing界面

(3) 点击Product→Archive进行归档，如图7-18所示，然后就可以点击Update to App Store...按钮来上传了，如图7-19所示。也可以将Archive包导出为Ad Hoc Deployment后，通过Application Loader来上传。

图7-18　Archive Information界面

图7-19 Archive Submiting界面

上传成功后，就可以将iTunes Connect中的应用提交审核了。当审核被拒绝的时候，苹果人员会发邮件给审核信息中填写的地址，描述对应的问题。除了要求应用正常运行外，还有很多内容以及政策上的要求，具体可以参考官网的审核指南。

第四部分
实战篇

- 第8章　企业内部通讯录应用开发
- 第9章　基于LBS的应用开发
- 第10章　豆搜App

第 8 章 企业内部通讯录应用开发

在前7章中，我们学习了React Native的语法、组件以及API。在这一章中，我们将进入实战环节，开发小型的App。进入一个新公司的时候，通常都会做的一件事就是加同事的微信、记电话号码。但是，我们很少去添加整个部门同事的号码，而仅仅添加自己组同事的号码。现实情况却是悲剧的，开个会需要同整个部门沟通，此时才发现电话号码太少。因此，能否开发一个部门内部的App供大家使用呢？在这一章中，我们开发一款部门通讯录App——百灵鸟。如果想直接看源码，可以直接访问GitHub：https://github.com/vczero/React-Native-App。

8.1 需求提出

手机上都有通讯录功能，使用它可以发送短信，拨打电话，而我们的"百灵鸟"App跟它很像。但是，有个区别是"百灵鸟"是"云通讯录"，管理员可以管理联系人信息，一个部门的同事共用的是一份联系人列表。

在开发前，我们需要明确这个App需要哪些功能。下面我们简单设计了"百灵鸟"的功能：

- ❑ 用户需要登录才能使用该App；
- ❑ 分组、分项目显示联系人列表；
- ❑ 可以显示联系人的组别、工作岗位、电话号码、邮箱；
- ❑ 可以给用户发送短信、邮件和拨打电话；
- ❑ 能够显示部门最近的公告消息；
- ❑ 能够搜索消息；
- ❑ 用户能修改自己的密码；
- ❑ 每个用户都是一个管理员，可以增加联系人、删除联系人、发布公告；
- ❑ 开发者的信息，提供bug反馈渠道。

为了方便开发，我们需要将以上功能通过流程图来表示，如图8-1所示。

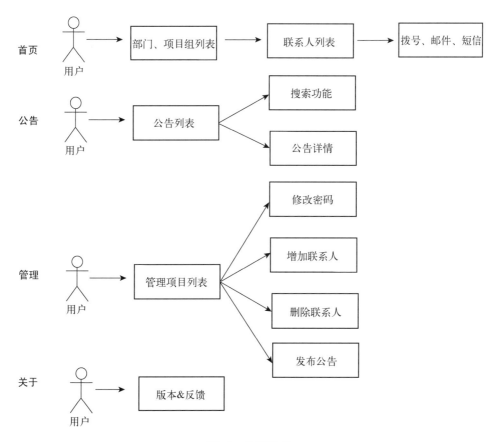

图8-1 功能需求

如图8-1所示,我们希望用户首先进入的页面是部门、项目组列表页,点击某一个组,就可以看到该组的联系人列表。然后可以拨打电话、发送邮件、短信给联系人。我们同时需要关注部门内部的动态,因此可以查看最近的公告和搜索历史公告。此外,还希望每一个内部员工是管理者,可以增加联系人、删除联系人、发布公告,同时可以修改自己的密码。

8.2 技术架构

当确定要开发一个应用的时候,随之而来的就是技术选型问题。这里开发的"百灵鸟"应用需要考虑到现实情况。一款内部App,没有必要驱动用户自行更新App。如果按照版本更新,反而会引起大家的反感,耽误大家的时间。因此,Web App和React Native是最佳选择。这里我们选择React Native,用户只需要安装一个壳即可。React Native App既能满足用户快速安装、稳定体验的特性,又能满足我们频繁更新通讯录的需要。既然是内部App,使用Node.js作为服务器端开发语言,不仅统一了开发语言,还轻便快速。

8.1节已经提出了4个功能需求，因此我们需要以需求驱动来完成整个App的设计。我们的整体技术架构如图8-2所示。

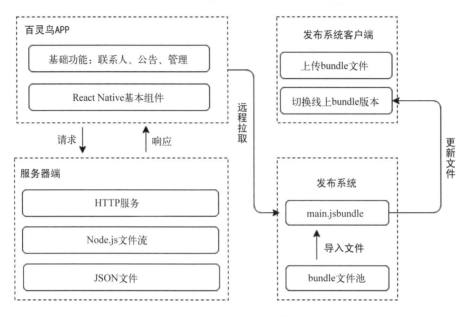

图8-2　整体技术架构

百灵鸟App整体分为3部分：客户端、服务器端以及发布系统。客户端部分以React Native基础组件为基础，开发联系人、公告、管理等功能。客户端调用的数据由Node.js服务器端提供。React Native jsbundle文件由发布系统提供。之所以将后端和发布系统独立，是因为二者更新频度不一样，所具有的功能不一样。服务器端我们采用了Node.js语言，将联系人数据、公告等数据都存入服务器端.json文件中，通过Node.js文件流进行数据的存储、删除、增加等操作。整个服务器端是一个轻量级的文件系统，暴露对文件内容修改的API。发布系统是为了独立发布，而不用跟服务器端混在一块，也没有必要混在一块。服务器端需要经常修改，但是发布系统往往比较稳定。我们在发布系统上选择一个版本，对应的App端就会更新内容。关于App的更新，可以参考第7章。关于jsbundle的更新，可以参考http://github.com/vczero/jsbundle，该项目是一个简易的jsbundle更新系统。

8.3　服务器端设计和开发

百灵鸟是一个小型通讯录App，我们没有必要把许多重武器加进去。我们不需要对用户的数据进行缓存，也没有必要为此添加一个数据库。为了快速实现高效开发，这里用的是基于文件系统的服务。也就是说，所有的通讯录信息都存在文件中。毕竟一个部门的通讯录信息比较少，也不是高频应用。

8.3.1 服务器端整体设计

很多时候，我们都有一个误区，那就是什么技术高大上就用什么技术，而不考虑应用的场景和团队的配置。百灵鸟App有必要使用数据库吗？我的答案是没有太大必要。因为一个部门就100人左右，用户的信息也就100条记录。完全没必要使用MySQL或者MongoDB。大家都知道服务器的内存是十分宝贵的。如果为了百来条记录开启一个数据库的服务器，那是对服务器内存极大的浪费。比如，MongoDB数据库服务启动后，立马就损失了上百兆的内存。那么，又有一个问题，那就是有没有必要使用缓存？我觉得没有必要。因为这是低频度的App，没有必要为用户数据做缓存。例如，如果增加Redis缓存，那么就需要维护缓存规则和命中条件。针对百灵鸟这款App而言，这是对人力的浪费。针对百来条记录，我们宁愿一次性添加到内存中，那样更快。因此，服务器端就是将所有数据都存储到json格式的文件中，所有的数据增加、删除、更新都是基于文件系统的。对于后期维护而言，可以减少很多不必要的麻烦。如果需要进行服务迁移，也会十分便利。后端的总体设计如图8-3所示。

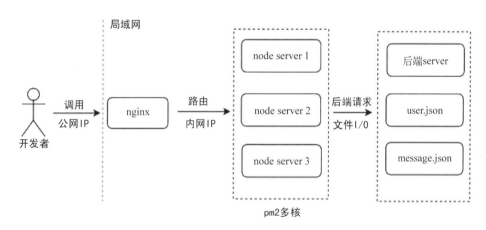

图8-3 服务器端设计

如图8-3所示，我们将Node.js服务和后端服务都部署在同一个局域网内。这样的话，网络请求的数据将会很快。比如，我们需要调用公司OA系统进行登录验证。图中user.json是用户的数据文件，message.json是公告消息的数据文件。

8.3.2 用户数据模型设计

我们没有使用数据库，使用的是文件存储。这里使用的是.json文件，其好处就是JSON对象的转化、存取十分方便。根据8.1节的需求，我们需要存储用户的姓名、加密密码、组别、电话、邮箱、标签等信息。因此，我们设计如下数据模型（单个员工信息）：

```
{
  "userid": "591F68A5-87E6-4A6C-810F-5A50953AE747",
```

```
    "username": "宋江",
    "password": "19bdec7440acd44c669240ed534fc2f6",
    "partment": "框架研发",
    "tel": "19008097890",
    "email": "test@126.com",
    "tag": "研发",
    "creater": "wlh",
    "token": "5C391E50-C160-4AFA-A4D5-19B315F5D357C34C0B51-3A68-4297-9C7E-A851B136D34F"
}
```

在上述代码中，各个属性的含义如下所示。

- **userid**：员工的唯一标识，便于我们删除用户。
- **username**：用户名。
- **password**：加密后的密码。
- **partment**：部门或者组别。
- **tel**：员工电话。
- **email**：员工邮箱。
- **tag**：标识员工序列，例如研发、产品。
- **creater**：创建者，该员工由谁添加。
- **token**：该员工的token，通过登录来换取新的token。token是权限的唯一标识。

既然设计好了员工模型，那么整个user.json应该是一个数组。该数组包含多个用户模型。user.json文件的内容如下：

```
[
    {
        "userid": "591F68A5-87E6-4A6C-810F-5A50953AE747",
        "username": "宋江",
        "password": "19bdec7440acd44c669240ed534fc2f6",
        "partment": "框架研发",
        "tel": "19008097890",
        "email": "test@126.com",
        "tag": "研发",
        "creater": "wlh",
        "token": "5C391E50-C160-4AFA-A4D5-19B315F5D357C34C0B51-3A68-4297-9C7E-A851B136D34F"
    },
    {
        "userid": "6998EED4-BB29-469E-B19A-DF4171B1FD12",
        "username": "卢俊义",
        "password": "19bdec7440acd44c669240ed534fc2f6",
        "partment": "监控研发",
        "tel": "19008097890",
        "email": "test@126.com",
        "tag": "研发",
        "creater": "wlh",
        "time": "2015-07-14T05:48:56.192Z",
        "token": "C0B668AB-A030-4A35-8A3C-422DEF4F9BE2undefined"
    },
```

```json
{
  "userid": "3240870A-538F-4A84-B00E-567E910A30ED",
  "username": "吴用",
  "password": "91266e54a7d08e3df4b26839ee946628",
  "partment": "框架产品",
  "tel": "19008097890",
  "email": "test@126.com",
  "tag": "产品",
  "creater": "wlh",
  "time": "2015-07-14T09:18:59.174Z",
  "token": "7722D6B3-8ACF-4C9B-8677-3326C14B7D9E5839849B-97AB-443D-B083-4236E550FAD2"
}
]
```

我们为每一个员工模型增加creater字段是十分必要的。因为我们App的权限是公开的，内部员工都能对数据进行修改，我们相信员工可以做到准确无误，但是做数据日志记录是必要的。用户的密码一定要加密加盐处理，这是对用户的保障。token的换取也要做到安全、可控。

8.3.3 公告数据模型设计

我们需要的不仅仅是通讯录功能，更希望可以发表一些状态信息。比如有新员工加入时，可以在公告里发表"欢迎致辞"。因此，我们需要设计公告的数据模型，其中应该包含用户名、用户ID、发表时间、消息内容等字段。

根据公告数据模型，我们的message.json文件应该是一个数组，该数组包含多个消息模型，其内容如下：

```json
[
  {
    "messageid": "3",
    "userid": "112222",
    "time": "2015-05-03",
    "username":"王**",
    "message": "欢迎小鹿同学加入携程框架部"
  },
  {
    "messageid": "4",
    "userid": "112222",
    "time": "2015-05-03",
    "username":"魏**",
    "message": "moles项目组周末一起去团建呗。"
  }
]
```

在上述代码中，消息模型中的各个字段解释如下。

❑ `messageid`：消息的ID。
❑ `userid`：用户ID。
❑ `time`：时间字符串，格式为yyyy-MM-dd。

- **username**：用户名。
- **message**：消息内容。

8.3.4 服务路由设计

此外，我们还希望从一个路由的URL串中看出该服务是干什么的，此时设计一套良好的服务API是必要的。因为"百灵鸟"App的功能比较少，所以这里设计的API比较简单。如果是大型的服务器端，可以参考RESTful风格设计API。我们设计的API URL如下：

```
//user API
//获取用户信息
app.post('/user/get', this.getUser);
//创建用户
app.post('/user/create', this.addUser);
//登录
app.post('/user/login', this.login);
//通过token登录
app.post('/user/login/token', this.loginByToken);
//更新密码
app.post('/user/password/update', this.updatePassword);
//删除用户
app.post('/user/delete', this.deleteUser);
//message API
//获取公告
app.post('/message/get', this.getMessage);
```

这里我们统一使用POST请求，当然也可以采用GET请求，比如获取用户信息时可以使用GET请求。

8.3.5 创建项目

后端的设计基本完成了，具体细节将会在开发中讲解，比如登录鉴权认证。在这一节中，我们开始搭建服务器端的项目。服务器端使用的是Node.js平台，使用的框架是express。我们按照以下步骤来完成项目的搭建（前提是已经安装好了Node.js）。

1. 创建项目

我们使用express创建项目，终端命令如下：

```
$ npm install -g express
$ npm install -g express-generator@4
$ express -e server
```

2. 搭建目录结构

良好的目录结构便于我们开发和管理项目，这里搭建的目录结构如图8-4所示。

图8-4 目录结构

在server下,我们新建了database目录,其中新建了两个.json文件。message.json是存储公告信息的文件,其内容目前是空数组,即[ ]。user.json是员工信息存储文件,其内容目前是空数组,即[ ]。routes是我们关注的路由服务目录。util.js模块包含一些工具方法,例如加密和GUID的生成等。routes.js是路由加载模块,主要将所有的模块统一加载到内存中。

3. 安装依赖

创建完项目后,我们需要安装相关的依赖。package.json文件中的dependencies是我们需要的库,相关代码如下:

```
{
  "name": "server",
  "version": "0.0.0",
  "private": true,
  "scripts": {
    "start": "node ./bin/www"
  },
  "dependencies": {
    "async": "^1.3.0",
    "body-parser": "~1.12.0",
    "cookie-parser": "~1.3.4",
    "debug": "~2.1.1",
    "ejs": "~2.3.1",
    "express": "~4.12.0",
    "morgan": "~1.5.1",
    "serve-favicon": "~2.2.0",
    "supervisor": "^0.7.1",
    "xss": "^0.2.7"
  }
}
```

然后在server目录下执行如下命令即可安装依赖:

```
$ npm install
```

8.3.6 app.js 入口文件

这里我们将app.js作为入口文件。因此，修改后的app.js文件的内容如下：

```js
var express = require('express');
var http = require('http');
var path = require('path');
var favicon = require('serve-favicon');
var logger = require('morgan');
var cookieParser = require('cookie-parser');
var bodyParser = require('body-parser');
var async = require('async');
var routes = require('./routes/routes');
var app = express();

app.set('port', 3000);
app.set('views', path.join(__dirname, 'views'));
app.set('view engine', 'ejs');

app.use(favicon(__dirname + '/public/favicon.ico'));
app.use(logger('dev'));
app.use(bodyParser.json());
app.use(bodyParser.urlencoded({ extended: true }));
app.use(cookieParser());
app.use(express.static(path.join(__dirname, 'public')));

var server = http.createServer(app);
server.listen(app.get('port'));

server.on('listening', function(){
  console.log('---listening on port: ' + app.get('port') +'---');
});

server.on('error', function(error){
  switch (error.code) {
    case 'EACCES':
      console.error(bind + '需要权限许可');
      process.exit(1);
      break;
    case 'EADDRINUSE':
      console.error(bind + '端口已被占用');
      process.exit(1);
      break;
    default:
      throw error;
  }
});

//加载路由
async.waterfall([
  function(callback){
    routes(app);
    callback(null);
```

```
    },
    function(){
      app.use(function(req, res, next) {
        var err = new Error('Not Found');
        err.status = 404;
        next(err);
      });

      if (app.get('env') === 'development') {
        app.use(function(err, req, res, next) {
          res.status(err.status || 500);
          res.render('404/error', {
            message: err.message,
            error: err
          });
        });
      }

      app.use(function(err, req, res, next) {
        res.status(err.status || 500);
        res.render('404/error', {
          message: err.message,
          error: {}
        });
      });
    }
]);
```

我们在app.js文件中加载了require('./routes/routes')模块，并且将express的实例作为参数传入了routes(app)。前面我们已经提到，routes模块的功能是将所有的服务加载到内存中。

8.3.7 加载服务模块到内存

server/routes/routes.js文件十分重要，通过它来加载服务模块，其代码如下：

```
var fs = require("fs");

module.exports = function(app){
  var FS_PATH_SERVICES = './routes/services/';
  var REQUIRE_PATH_SERVICES = './services/';

  fs.readdir(FS_PATH_SERVICES, function(err, list){
    if(err){
      throw '没有找到该文件夹，请检查……'
    }
    for (var e; list.length && (e = list.shift());){
      var service = require(REQUIRE_PATH_SERVICES + e);
      service.init && service.init(app);
    }
  });
};
```

我们使用require加载了server/routes/services/目录下的所有模块。并且调用每个模块的init方法。这样做的一个好处是可以将对象封装。例如，我们可以添加test.js。init函数可以使用app参数挂载路由，所有的请求路由都集中在该函数中。这样做的好处是，一个模块的路由和功能都在一个文件中，便于管理和维护。test.js文件的内容如下：

```javascript
var Test = {
  init: function(app){
    app.get('/test/test', this.doTest);
    app.get('/test/show', this.doShow);
  },

  doTest: function(req, res){
    res.send({
      status: 1,
      info: '测试服务doTest'
    });
  },

  doShow: function(req, res){
    res.json({
      status: 1,
      info: '测试服务doShow'
    });
  }
};

module.exports = Test;
```

我们可以给message.js和user.js写个简单模块，然后使用node app.js启动服务，此时便可以看到服务返回的数据了。例如，通过浏览器访问http://localhost:3000/test/show，可以得到如下结果：

```
{"status":1,"info":"测试服务doShow"}
```

这说明我们的服务模块加载成功。

8.3.8 工具类开发

我们已经打通了服务模块的加载流程，现在需要提前开发一些工具方法。我们在util.js文件中增加如下代码：

```javascript
var crypto = require('crypto');
module.exports = {
  guid: function() {
    return 'xxxxxxxx-xxxx-4xxx-yxxx-xxxxxxxxxxxx'.replace(/[xy]/g, function(c) {
      var r = Math.random() * 16 | 0,
          v = c == 'x' ? r : (r & 0x3 | 0x8);
      return v.toString(16);
    }).toUpperCase();
  },
```

```
    md5: function(password){
      var md5 = crypto.createHash('md5');
      var salt = '(!%$88hs@gophs*)#sassb9';
      var newPwd = md5.update(password + salt).digest('hex');
      return newPwd;
    },
    getKey: function(){
      return 'HSHHSGSGGSTWSYWSYUSUWSHWBS-REACT-NATIVE';
    }
};
```

该工具类主要有3个函数：guid用于生成随机ID（源自Robert Kieffer），md5函数主要用于加密用户密码，getKey用于返回鉴权key值。

8.3.9 用户信息接口

我们已经开发好了routes模块加载器，同时也开发好了工具类，下一步就是实现服务。我们在server/routes/services/user.js文件中开发用户信息相关的服务。

1. 获取用户信息

在init函数中增加路由app.post('/user/get', this.getUser)。getUser是获取信息服务，这里需要传入使用鉴权key和部门，才能查询出数据。这里，我们读取的是user.json文件。当查询到符合条件的数据时，我们删除密码并返回数据。

2. 添加用户

在init函数中增加路由app.post('/user/create', this.addUser)，其中addUser是增加用户服务。我们需要传入用户名、密码、电话、邮箱、部门组别、标签（产品、研发和测试）、添加者等信息才能构建一个完成的用户对象。首先，我们从user.json中读取数据并序列化成JSON对象，然后在JSON对象上推送数据。

3. 用户登录

在init函数中增加路由app.post('/user/login', this.login)，其中login是登录服务，该服务需要传入用户的邮箱、密码和设备ID。我们使用工具类的guid方法和设备ID生成了token。这里需要删除密码并返回该用户相关的数据。

4. 使用token登录

在init函数中增加路由app.post('/user/login/token', this.loginByToken)，其中loginByToken是根据token登录的服务。

5. 更新用户密码

在init函数中增加路由app.post('/user/password/update', this.updatePassword)，其中

updatePassword是更新密码服务。

6. 删除用户服务

在init函数中增加路由app.post('/user/delete', this.deleteUser)，其中deleteUser是删除用户服务。需要传入的参数是被删除者的邮箱和删除者的token。

user.js文件的完整代码如下所示：

```javascript
var fs = require('fs');
var util = require('./../util');
var USER_PATH = './database/user.json';

var User = {

  init: function(app){
    app.post('/user/get', this.getUser);
    app.post('/user/create', this.addUser);
    app.post('/user/login', this.login);
    app.post('/user/login/token', this.loginByToken);
    app.post('/user/password/update', this.updatePassword);
    app.post('/user/delete', this.deleteUser);
  },

  //获取用户信息
  getUser: function(req, res){
    var key = req.param('key');
    var partment = req.param('partment');
    if(key !== util.getKey()){
      return res.send({
        status: 0,
        data: '使用了没有鉴权的key'
      });
    }
    fs.readFile(USER_PATH, function(err, data){
      if(!err){
        try{
          var obj = JSON.parse(data);
          var newObj = [];
          for(var i in obj){
            if(obj[i].partment === partment){
              delete obj[i]['password'];
              newObj.push(obj[i]);
            }
          }
          return res.send({
            status: 1,
            data: newObj
          });
        }catch(e){
          return res.send({
            status: 0,
            err: e
          });
```

```javascript
      }
    }
    return res.send({
      status: 0,
      err: err
    });
  });
},

//添加用户
addUser: function(req, res){
  var username = req.param('username');
  var password = util.md5(req.param('password'));
  var tel = req.param('tel');
  var email = req.param('email');
  var partment =  req.param('partment');
  var tag = req.param('tag');
  var creater = req.param('creater') || '';

  if(!username || !password || !tel || !email || !partment || !tag || !creater){
    return res.send({
      status: 0,
      data: '缺少必要参数'
    });
  }

  try{
    var content = JSON.parse(fs.readFileSync(USER_PATH));
    var obj = {
      "userid": util.guid(),
      "username": username,
      "password": password,
      "partment": partment,
      "tel": tel,
      "email": email,
      "tag": tag,
      "creater": creater,
      "time": new Date(),
      "token": ''
    };
    content.push(obj);
    //更新文件
    fs.writeFileSync(USER_PATH, JSON.stringify(content));
    delete obj.password;
    return res.send({
      status: 1,
      data: obj
    });
  }catch(e){
    return res.send({
      status: 0,
      err: e
    });
  }
}
```

```javascript
    },
    //用户登录
    login: function(req, res){
      var email = req.param('email');
      var password = util.md5(req.param('password'));
      var deviceId = req.param('deviceId');
      var token = util.guid() + deviceId;
      var content = JSON.parse(fs.readFileSync(USER_PATH).toString());
      for(var i in content){
        //验证通过
        if(content[i].email === email && content[i].password === password){
          content[i]['token'] = token;
          //写入到文件中
          console.log(content[i]);
          fs.writeFileSync(USER_PATH, JSON.stringify(content));
          //删除密码
          delete content[i].password;
          return res.send({
            status: 1,
            data: content[i]
          });
        }
      }

      return res.send({
        status: 0,
        data:'用户名或者密码错误'
      });
    },

    //通过token登录
    loginByToken: function(req, res){
      var token = req.param('token');
      var content = JSON.parse(fs.readFileSync(USER_PATH));

      for(var i in content){
        if(token === content[i].token){
          delete content[i].password;
          return res.send({
            status:1,
            data: content[i]
          });
        }
      }

      return res.send({
        status: 0,
        info: 'token失效'
      });
    },

    //用户修改密码
    updatePassword: function(req, res){
      var token = req.param('token');
```

```javascript
      var oldPassword = util.md5(req.param('oldPassword'));
      var password = util.md5(req.param('password'));

      var content = JSON.parse(fs.readFileSync(USER_PATH));
      for(var i in content){
        if(token === content[i].token && oldPassword === content[i].password){
          content[i].password = password;
          //写入到文件中
          fs.writeFileSync(USER_PATH, JSON.stringify(content));
          return res.send({
            status: 1,
            data: '更新成功'
          });
        }
      }

      return res.send({
        status: 0,
        data: '更新失败,没有找到该用户或者初始密码错误'
      });
    },

    //删除用户
    deleteUser: function(req, res) {
      var token = req.param('token');
      var email = req.param('email');

      var content = JSON.parse(fs.readFileSync(USER_PATH));
      for (var i in content) {
        if (token === content[i].token) {
          //遍历查找需要删除的用户
          for (var j in content) {
            if (content[j].email === email) {
              content.splice(j, 1);
              //写入到文件中
              fs.writeFileSync(USER_PATH, JSON.stringify(content));
              return res.send({
                status: 1,
                info: content,
                data: '删除成功'
              });
            }
          }
        }
      }
      return res.send({
        status: 0,
        err: '删除失败,没有找到该用户或者用户鉴权错误'
      });
    }
};

module.exports = User;
```

8.3.10 公告消息接口

我们在server/routes/services/message.js文件中添加如下代码:

```javascript
var fs = require('fs');
var util = require('./../util');
var MESSAGE_PATH = './database/message.json';
var USER_PATH = './database/user.json';

var Message = {
  init: function(app){
    app.post('/message/get', this.getMessage);
    app.post('/message/add', this.addMessage);
  },

  //获取公告消息
  getMessage: function(req, res){
    var key = req.param('key');
    if(key !== util.getKey()){
      return res.send({
        status: 0,
        data: '使用了没有鉴权的key'
      });
    }
    fs.readFile(MESSAGE_PATH, function(err, data){
      if(!err){
        try{
          var obj = JSON.parse(data);
          return res.send({
            status: 1,
            data: obj
          });
        }catch(e){
          return res.send({
            status: 0,
            err: e
          });
        }
      }

      return res.send({
        status: 0,
        err: err
      });
    });
  },

  //增加公告消息
  addMessage: function(req, res){
    var token = req.param('token');
    var message = req.param('message');
    if(!token || !message){
      return res.send({
```

```javascript
          status: 0,
          err: 'token或者message不能为空'
        });
      }
      //根据token查询
      fs.readFile(USER_PATH, function(err, data){
        if(err){
          return res.send({
            status: 0,
            err: err
          });
        }

        try{
          var obj = JSON.parse(data);
          for(var i in obj){
            if(obj[i].token === token){
              //增加信息
              var msgObj = JSON.parse(fs.readFileSync(MESSAGE_PATH));
              msgObj.push({
                messageid: util.guid(),
                userid: obj[i].userid,
                username: obj[i].username,
                time: new Date().getFullYear() + '-' 
                    + (parseInt(new Date().getMonth()) + 1) + '-' + new Date().getDate(),
                message: message
              });

              fs.writeFileSync(MESSAGE_PATH, JSON.stringify(msgObj));
              return res.send({
                status: 1
              });
            }
          }
          return res.send({
            status: 0,
            err: 'token认证失败'
          });
        }catch(e){
          return res.send({
            status: 0,
            err: e
          });
        }
      });
    }
  };

module.exports = Message;
```

在上述代码中，getMessage方法用于获取message.json文件中的数据，addMessage方法用于向message.json文件中增加数据。

8.3.11 建议

这里服务器端的代码只是作为实例，还有很多需要完善的地方，例如：

- 增加XSS过滤，防止恶意表单；
- 增加鉴权认证，保证数据接口安全；
- 增加日志，记录用户登录和使用情况；
- 增加返回结果条数控制等。

服务器端的代码托管在 GitHub 上（详见 https://github.com/vczero/React-Native-App/tree/master/address_book/server），欢迎大家完善代码。

8.4 客户端设计和开发

现在，我们需要开发React Native iOS客户端。

8.4.1 客户端设计

关于客户端的开发，我们采用的是React Native技术。React Native提倡的是组件化，因此，我们需要重点考虑组件的颗粒度。比如有人将登录页面作为一个组件，他的理由是整个登录页面是唯一的，其他组件只是调用登录页面而已。而有的人将input输入框单独出一个组件，他认为搜索输入框、登录输入框都是同一类。这里，我更倾向于前者，因为基于业务的组件跟业务绑在一块更好，方便后期代码的更新和维护。

1. 客户端总体设计

其实每一个功能都有很多技术方向的选择，但是选择最贴合当前条件和应用的才是最好的。比如，有时候为了实现一个"加入购物车"的动画特效，去加载一个动画库是不值得的。如果整个App有十多处动画，并且对动画要求较高，但是又没有时间去实现一套动画，那么加载开源的动画库是个明智的选择。这里，百灵鸟是一个简单的App，我们需要做到的是信息正确。客户端整体设计如图8-5所示。

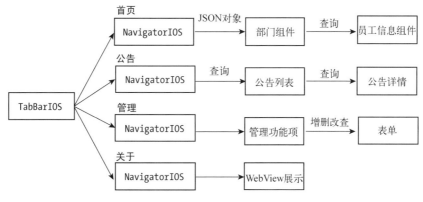

图8-5　客户端总体设计

可以看出，路由的深度是2，所以对用户来说，很容易操作。TabBarIOS中设置4个选项卡，每个选项卡都是独立的NavigatorIOS。也就是说，每个功能模块独立路由。

2. 目录结构

整个客户端的设计已经做好，现在我们需要规划开发目录，如图8-6所示。

图8-6　开发目录

在开发目录中，index.ios.js是客户端程序的入口文件，address_book/views是所有组件存放目录。各文件的含义如图8-7所示。

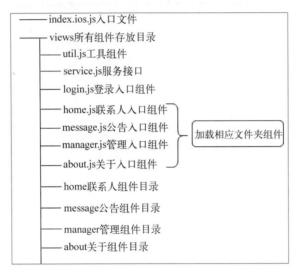

图8-7 组件目录

8.4.2 工具组件和服务

util.js文件是工具模块，封装了获取屏幕尺寸和POST请求，具体的代码如下：

```
var React = require('react-native');
var Dimensions = require('Dimensions');

var {
  PixelRatio
} = React;

var Util = {

  //单位像素
  pixel: 1 / PixelRatio.get(),
  //屏幕尺寸
  size: {
    width: Dimensions.get('window').width,
    height: Dimensions.get('window').height
  },

  //POST请求
  post: function (url, data, callback) {
    var fetchOptions = {
      method: 'POST',
      headers: {
```

```
        'Accept': 'application/json',
        'Content-Type': 'application/json'
      },
      body: JSON.stringify(data)
    };

    fetch(url, fetchOptions)
    .then((response) => response.text())
    .then((responseText) => {
      callback(JSON.parse(responseText));
    });
  },
  //key
  key: 'HSHHSGSGGSTWSYWSYUSUWSHWBS-REACT-NATIVE'

};
module.exports = Util;
```

这里我们引用了Dimensions API来获取屏幕的宽度和高度，使用PixelRatio.get()来获取屏幕像素比，同时将fetch方法封装成了post方法，便于我们发出POST请求。

为了更好地管理服务的URL，我们将服务URL都集中到一个文件中，即service.js，该文件的内容如下：

```
var Service = {
  host:'http://127.0.0.1:3000',
  login: '/user/login',
  loginByToken: '/user/login/token',
  getUser: '/user/get',
  createUser: '/user/create',
  getMessage: '/message/get',
  addMessage: '/message/add',
  updatePassword: '/user/password/update',
  deleteUser: '/user/delete'
};

module.exports = Service;
```

这样，我们只用修改host就可以打包生产了。

8.4.3 添加依赖库

我们的项目需要获取设备的ID，因此需要加载RCTAdSupport项目。右击Libraries，选择Add Files to "address_book"...，如图8-8所示。

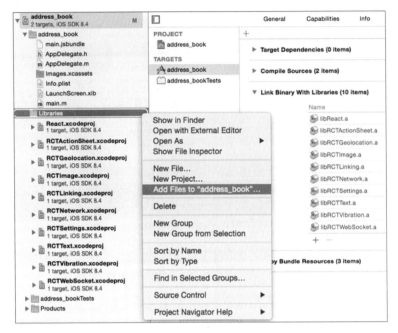

图8-8 添加类库

此时会打开项目选择对话框，从中选择address_book/node_modules/react-native/Libraries/AdSupport/RCTAdSupport.xcodeproj文件，如图8-9所示。

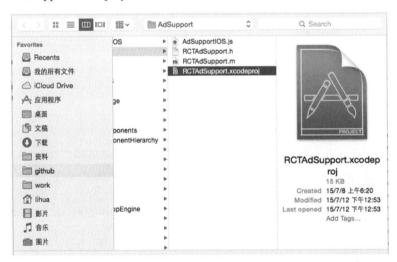

图8-9 添加RCTAdSupport.xcodeproj文件

添加完项目文件后，我们需要添加静态库。选中address_book项目，选中Build Phases，然后展开Link Binary With Libraries节点，如图8-10所示。

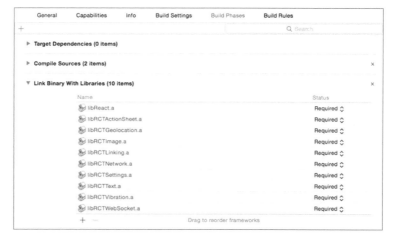

图8-10　展开Link Binary With Libraries节点

点击+图标，添加静态库文件。如图8-11所示，我们选中libRCTAdSupport.a。

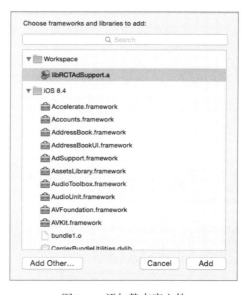

图8-11　添加静态库文件

8.4.4　程序入口和登录

　　index.ios.js是React Native的入口文件，我们可以在该入口文件中装载路由配置、全局变量等。在"百灵鸟"应用中，我们需要加载联系人、公告消息、信息管理、关于入口组件，具体代码如下：

```
var React = require('react-native');
var AdSupportIOS = require('AdSupportIOS');
var Home = require('./views/home');
var About = require('./views/about');
var Manager = require('./views/manager');
var Message = require('./views/message');
var Util = require('./views/util');
var Service = require('./views/service');
```

这里我们引入了AdSupportIOS组件,这是为了获取设备的ID。AdSupportIOS的用法如下:

```
AdSupportIOS.getAdvertisingTrackingEnabled(function(){
    AdSupportIOS.getAdvertisingId(function(deviceId){
        //TODO: deviceId即设备ID,我们可以将设备ID作为用户登录的token的一部分
    });
});
```

我们可以使用TabBarIOS和TabBarIOS.Item组件来加载不同的功能组件。当用户选中某个item时(即TabBarIOS.Item的selected为true时),会展现相应的视图,具体的代码如下:

```
<View style={this.state.showIndex}>
  <TabBarIOS barTintColor="#FFF">
    <TabBarIOS.Item
      icon={require('image!phone_s')}
      title="首页"
      selected={this.state.selectedTab === 'home'}
      onPress={this._selectTab.bind(this, 'home')}
      >
      {this._addNavigator(Home, '主页')}
    </TabBarIOS.Item>

    <TabBarIOS.Item
      title="公告"
      icon={require('image!gonggao')}
      selected={this.state.selectedTab === 'message'}
      onPress={this._selectTab.bind(this, 'message')}
      >
      {this._addNavigator(Message, '公告')}
    </TabBarIOS.Item>

    <TabBarIOS.Item
      title="管理"
      icon={require('image!manager')}
      selected={this.state.selectedTab === 'manager'}
      onPress={this._selectTab.bind(this, 'manager')}
      >
      {this._addNavigator(Manager, '管理')}
    </TabBarIOS.Item>

    <TabBarIOS.Item
      title="关于"
      icon={require('image!about')}
      selected={this.state.selectedTab === 'about'}
```

```
                onPress={this._selectTab.bind(this, 'about')}
            >
                {this._addNavigator(About, '关于')}
            </TabBarIOS.Item>
        </TabBarIOS>
    </View>
```

同时，在当前视图中增加了登录页面。如果用户登录了，则直接跳转到首页，如果没有登录则显示登录组件。我们在AsyncStorage中存储了用户的token，如果AsyncStorage不存在token或者token失效，则表明用户需要登录。index.ios.js文件的完整代码如下所示：

```
'use strict';
var React = require('react-native');
var AdSupportIOS = require('AdSupportIOS');
var Home = require('./views/home');
var About = require('./views/about');
var Manager = require('./views/manager');
var Message = require('./views/message');
var Util = require('./views/util');
var Service = require('./views/service');

var {
  StyleSheet,
  View,
  TabBarIOS,
  Text,
  NavigatorIOS,
  AppRegistry,
  Image,
  TextInput,
  StatusBarIOS,
  ScrollView,
  TouchableHighlight,
  ActivityIndicatorIOS,
  AlertIOS,
  AsyncStorage,
} = React;

StatusBarIOS.setStyle('light-content');
var Address = React.createClass({
  statics: {
    title: '主页',
    description: '选项卡'
  },

  getInitialState: function(){
    return {
      selectedTab: 'home',
      showIndex: {
        height:0,
        opacity:0
      },
      showLogin:{
```

```
          flex:1,
          opacity:1
        },
        isLoadingShow: false
      };
    },

    componentDidMount: function(){
      var that = this;
      AsyncStorage.getItem('token', function(err, token){
        if(!err && token){
          var path = Service.host + Service.loginByToken;
          Util.post(path, {
            token: token
          },function(data){
            if(data.status){
              that.setState({
                showLogin: {
                  height:0,
                  width:0,
                  flex:0,
                },
                showIndex:{
                  flex:1,
                  opacity:1
                },
                isLoadingShow: false
              });
            }
          });
        }else{
          that.setState({
            showIndex: {
              height:0,
              opacity:0
            },
            showLogin:{
              flex:1,
              opacity:1
            },
            isLoadingShow: false
          });
        }
      });

      var path = Service.host + Service.getMessage;
      var that = this;
      Util.post(path, {
        key: Util.key
      }, function(data){
        that.setState({
          data: data
        });
      });
    },
```

```
_selectTab: function(tabName){
  this.setState({
    selectedTab: tabName
  });
},
_addNavigator: function(component, title){
  var data = null;
  if(title === '公告'){
    data = this.state.data;
  }
  return <NavigatorIOS
    style={{flex:1}}
    barTintColor='#007AFF'
    titleTextColor="#fff"
    tintColor="#fff"
    translucent={false}
    initialRoute={{
      component: component,
      title: title,
      passProps:{
        data: data
      }
    }}
  />;
},
_getEmail: function(val){
  var email = val;
  this.setState({
    email: email
  });
},
_getPassword: function(val){
  var password = val;
  this.setState({
    password: password
  });
},
_login: function(){
  var email = this.state.email;
  var password = this.state.password;
  var path = Service.host + Service.login;
  var that = this;

  //隐藏登录页并且加载loading效果
  that.setState({
    showLogin: {
      height:0,
      width:0,
      flex:0,
    },
```

```
            isLoadingShow: true
        });
        AdSupportIOS.getAdvertisingTrackingEnabled(function(){
            AdSupportIOS.getAdvertisingId(function(deviceId){
                Util.post(path, {
                    email: email,
                    password: password,
                    deviceId: deviceId,
                }, function(data){
                    if(data.status){
                        var user = data.data;
                        //加入数据到本地
                        AsyncStorage.multiSet([
                            ['username', user.username],
                            ['token', user.token],
                            ['userid', user.userid],
                            ['email', user.email],
                            ['tel', user.tel],
                            ['partment', user.partment],
                            ['tag', user.tag],
                        ], function(err){
                            if(!err){
                                that.setState({
                                    showLogin: {
                                        height:0,
                                        width:0,
                                        flex:0,
                                    },
                                    showIndex:{
                                        flex:1,
                                        opacity:1
                                    },
                                    isLoadingShow: false
                                });
                            }
                        });

                    }else{
                        AlertIOS.alert('登录', '用户名或者密码错误');
                        that.setState({
                            showLogin: {
                                flex:1,
                                opacity:1
                            },
                            showIndex:{
                                height:0,
                                width:0,
                                flex:0,
                            },
                            isLoadingShow: false
                        });
                    }
                });
            }, function(){
                AlertIOS.alert('设置','无法获取设备唯一标识');
```

```
        });
    }, function(){
        AlertIOS.alert('设置','无法获取设备唯一标识,请关闭设置→隐私→广告→限制广告跟踪');
    });
},

render: function(){
    return(
        <View style={{flex:1}}>
            {this.state.isLoadingShow ?
                <View style={{flex:1, justifyContent:'center', alignItems:'center'}}>
                    <ActivityIndicatorIOS size="small" color="#268DFF"></ActivityIndicatorIOS>
                </View>:null
            }
            {!this.state.isLoadingShow ?
                <View style={this.state.showIndex}>
                    <TabBarIOS barTintColor="#FFF">
                        <TabBarIOS.Item
                            icon={require('image!phone_s')}
                            title="首页"
                            selected={this.state.selectedTab === 'home'}
                            onPress={this._selectTab.bind(this, 'home')}
                            >
                            {this._addNavigator(Home, '主页')}
                        </TabBarIOS.Item>

                        <TabBarIOS.Item
                            title="公告"
                            icon={require('image!gonggao')}
                            selected={this.state.selectedTab === 'message'}
                            onPress={this._selectTab.bind(this, 'message')}
                            >
                            {this._addNavigator(Message, '公告')}
                        </TabBarIOS.Item>

                        <TabBarIOS.Item
                            title="管理"
                            icon={require('image!manager')}
                            selected={this.state.selectedTab === 'manager'}
                            onPress={this._selectTab.bind(this, 'manager')}
                            >
                            {this._addNavigator(Manager, '管理')}
                        </TabBarIOS.Item>

                        <TabBarIOS.Item
                            title="关于"
                            icon={require('image!about')}
                            selected={this.state.selectedTab === 'about'}
                            onPress={this._selectTab.bind(this, 'about')}
                            >
                            {this._addNavigator(About, '关于')}
                        </TabBarIOS.Item>
                    </TabBarIOS>
                </View> : null
            }
```

```jsx
        <ScrollView style={[this.state.showLogin]}>
            <View style={styles.container}>
               <View>
                 <Image style={styles.logo} source={require('image!logo')}></Image>
               </View>

               <View style={styles.inputRow}>
                 <Text>邮箱</Text><TextInput style={styles.input}
                    placeholder="请输入邮箱" onChangeText={this._getEmail}/>
               </View>
               <View style={styles.inputRow}>
                 <Text>密码</Text><TextInput style={styles.input}
                    placeholder="请输入密码" password={true} onChangeText={this._getPassword}/>
               </View>

               <View>
                 <TouchableHighlight underlayColor="#fff" style={styles.btn}
                    onPress={this._login}>
                    <Text style={{color:'#fff'}}>登录</Text>
                 </TouchableHighlight>
               </View>
            </View>
        </ScrollView>

      </View>
    );
  }

});

var styles = StyleSheet.create({
  container:{
    marginTop:50,
    alignItems:'center',
  },
  logo:{
    width:100,
    height:100,
    resizeMode: Image.resizeMode.contain
  },
  inputRow:{
    flexDirection:'row',
    alignItems:'center',
    justifyContent: 'center',
    marginBottom:10,
  },
  input:{
    marginLeft:10,
    width:220,
    borderWidth:Util.pixel,
    height:35,
    paddingLeft:8,
    borderRadius:5,
    borderColor:'#ccc'
  },
```

```
  btn:{
    marginTop:10,
    width:80,
    height:35,
    backgroundColor:'#3BC1FF',
    justifyContent:'center',
    alignItems:'center',
    borderRadius: 4,
  }
});

AppRegistry.registerComponent('address_book', () => Address);
```

在上述代码中,我们需要注意以下3点。

- 在componentDidMount中,首先从AsyncStorage获取token,然后通过向服务器端发送请求,判断用户的token是否有效。如果验证失败,则显示登录组件。
- 同时,我们给每个TabBarIOS.Item组件分配了一个NavigatorIOS组件,作为该功能模块的路由组件。
- 在登录函数_login中,我们需要使用AsyncStorage.multiSet设置多个字段的数据,同时生成以设备ID为基础的用户token。

此时首页已经开发完成,TabBar使用的icon可以到https://github.com/vczero/React-Native-App/tree/master/address_book/iOS/Images.xcassets下载。下载完成后,将该目录下的所有图片文件拖进Xcode的Images.xcassets即可。完成后的登录页面如图8-12所示。

图8-12 登录页

8.4.5 联系人列表

我们默认选中的是主页（selectedTab='home'），所以登录后默认加载的是联系人页面。也就是说，在index.ios.js中_addNavigator函数传递的是Home组件。首先，我们需要完成部门组别的列表，如图8-13所示。

图8-13　部门组别列表

首先，需要将渲染的组别信息构建成JSON对象，然后将对象的数组渲染成一个个矩形方块。我们在address_book/views/home.js中添加如下代码：

```
var React = require('react-native');
var Util = require('./util');
var ItemBlock = require('./home/itemblock');

var {
  View,
  Text,
  ScrollView,
  StyleSheet,
  TouchableHighlight,
} = React;

var Home = React.createClass({
  getInitialState: function(){
    //减去paddingLeft && paddingRight && space
    var width = Math.floor(((Util.size.width - 20) - 50) / 4);
```

```
      var items = [
        {
          title: '研发',
          partment: '框架研发',
          color: '#126AFF',
        },
        {
          title: '研发',
          partment: 'BU研发',
          color: '#FFD600',
        },
        {
          title: '产品',
          partment: '公共产品',
          color: '#F80728',
        },
        {
          title: '产品',
          partment: 'BU产品',
          color: '#05C147',
        },
        {
          title: '产品',
          partment: '启明星',
          color: '#FF4EB9',
        },
        {
          title: '项目',
          partment: '项目管理',
          color: '#EE810D',
        }
      ];

      return {
        items: items,
        width: width
      };
    },

    render: function(){
      var Items1 = [];
      var Items2 = [];
      var items = this.state.items;

      for(var i = 0; i < 4; i++){
        Items1.push(
          <ItemBlock
            title={items[i].title}
            partment={items[i].partment}
            width={this.state.width}
            color={items[i].color}
            nav={this.props.navigator}
          />
        );
```

```
            }

            for(var i = 4; i < items.length; i++){
                Items2.push(
                    <ItemBlock
                        title={items[i].title}
                        partment={items[i].partment}
                        width={this.state.width}
                        color={items[i].color}
                        nav={this.props.navigator}
                    />
                );
            }

            return (
                <ScrollView style={styles.container}>
                    <View style={styles.itemRow}>
                        {Items1}
                    </View>
                    <View style={styles.itemRow}>
                        {Items2}
                    </View>

                </ScrollView>
            );
        }
});

var styles = StyleSheet.create({
    container:{
        flex:1,
        padding:10,
    },
    itemRow:{
        flexDirection:'row',
        marginBottom:20,
    }
});

module.exports = Home;
```

从上面的代码可以看出，我们加载了ItemBlock组件。因为这里将每个部门的矩形方块抽象成一个组件，即address_book/views/home/itemblock.js。ItemBlock组件的代码如下：

```
var React = require('react-native');
var Address = require('./address');
var Service = require('./../service');
var Util = require('../util');

var {
    View,
    Text,
    StyleSheet,
```

```
  TouchableHighlight,
} = React;

//每个单项组件
var ItemBlock = React.createClass({
  render: function(){
    var size ={
      width: parseInt(this.props.width),
      height: parseInt(this.props.width),
      backgroundColor: this.props.color,
    };
    return (
      <TouchableHighlight underlayColor="#fff" onPress={this._loadPage}>
        <View style={[styles.itemBlock, size]}>
          <View>
            <Text style={styles.font18}>{this.props.title}</Text>
          </View>
          <View>
            <Text style={styles.font10}>{this.props.partment}</Text>
          </View>
        </View>
      </TouchableHighlight>
    );
  },
  //加载页面
  _loadPage: function(e){
    var nav = this.props.nav;
    var key = Util.key;
    var partment = this.props.partment;
    var path = Service.host + Service.getUser;

    Util.post(path, {
      key: key,
      partment : partment
    }, function(data){
      nav.push({
        title: this.props.tag,
        component: Address,
        passProps:{
          data: data
        }
      });
    }.bind(this));

  }
});

var styles = StyleSheet.create({
  itemBlock:{
    justifyContent:'center',
    alignItems:'center',
    borderRadius:5,
    marginLeft:10,
  },
```

```
        font18:{
          color:'#fff',
          fontSize:18,
          fontWeight:'500',
        },
        font10:{
          color:'#fff',
          fontSize:10,
        },
});

module.exports = ItemBlock;
```

ItemBlock组件主要有5个属性供外部组件调用传递：width、height、backgroundColor、title和partment。这些属性都由Home组件的JSON对象传递。为了加载联系人列表，我们需要在点击某一个方块时加载相应的部门组别的联系人列表。在加载联系人列表前，我们需要准备好联系人的数据。因此，这里在成功调用Service.getUser服务后，我们才将新的路由推送到navigator中去。这里我们又要加载一个新的组件Address，这是因为我们需要根据联系人数据渲染页面。我们需要完成的效果如图8-14所示。

图8-14　联系人列表

因此，我们在address_book/views/home/address.js文件中添加如下代码：

```
var React = require('react-native');
var Util = require('../util');
var ActionSheetIOS = require('ActionSheetIOS');
var Service = require('./../service');
var {
  View,
  Text,
  ScrollView,
  StyleSheet,
  Image,
  TouchableHighlight,
  LinkingIOS,
  AlertIOS,
} = React;

var Address = React.createClass({
  render: function(){
    var view = [];
    var items = this.props.data.status? this.props.data.data: [];
    var colors = ['#E20079', '#FFD602', '#25BFFE', '#F90000',
      '#04E246', '#04E246', '#00AFC9'];
    var color = {
      backgroundColor: colors[parseInt(Math.random()*7)]
    };
    for(var i in items){
      view.push(
        <View style={styles.row}>
          <View style={[styles.text, color]}>
            <Text style={{fontSize:25, color:'#fff', fontWeight:'bold'}}>
              {items[i].username.substr(0, 1) || '未'}
            </Text>
          </View>
          <View style={styles.part}>
            <Text>
              {items[i].username}
            </Text>
            <Text style={styles.unColor}>
              {(items[i].partment||'') + '部-' + (items[i].tag||'') + '人员'}
            </Text>
          </View>
          <View style={{flex:1}}>
            <TouchableHighlight underlayColor="#fff"
              onPress={this.showActionSheet.bind(this, items[i].tel,
                items[i].email, items[i].username)}>
              <Text style={styles.link}>
                {items[i].tel}
              </Text>
            </TouchableHighlight>
            <TouchableHighlight underlayColor="#fff"
              onPress={this.showActionSheet.bind(this, items[i].tel,
                items[i].email, items[i].username)}>
              <Text style={styles.link}>
                {items[i].email}
```

```
                </Text>
              </TouchableHighlight>
            </View>
          </View>
        );
      }
      return (
        <ScrollView>
          {view}
        </ScrollView>
      );
    },
    showActionSheet(tel, email, name) {
      var options = [];
      options.push('拨打电话给：' + name);
      options.push('发送短信给：' + name);
      options.push('发送邮件给：' + name);
      options.push('取消');

      var events = [];
      events.push(function(){
        LinkingIOS.openURL('tel://' + tel);
      });
      events.push(function(){
        LinkingIOS.openURL('sms://' + tel);
      });
      events.push(function(){
        LinkingIOS.openURL('mailto://' + email);
      });

      ActionSheetIOS.showActionSheetWithOptions({
          options: options,
          cancelButtonIndex: options.length - 1 ,
        },
        function(index){
          events[index] && events[index]();
        }
      );
    }
});

var styles = StyleSheet.create({
  row:{
    height:80,
    borderBottomWidth: Util.pixel,
    borderBottomColor:'#ccc',
    flexDirection:'row',
    alignItems:'center'
  },
  text:{
    width:50,
    height:50,
```

```
    borderRadius:4,
    marginLeft:10,
    alignItems:'center',
    justifyContent:'center',
    backgroundColor: '#E30082',
  },
  part:{
    marginLeft:5,
    flex:1,
  },
  link:{
    color:'#1BB7FF',
    marginTop:2,
  },
  unColor:{
    color: '#575656',
    marginTop:8,
    fontSize:12,
  }
});

module.exports = Address;
```

在Address组件中，我们使用LinkingIOS组件来调用iOS系统的拨号、发短信、发邮件等功能。点击用户的电话和邮箱，效果如图8-15所示。

图8-15　点击效果

8.4.6 公告功能

公告功能主要有两个页面：列表页和详情页。address_book/views/message.js文件是我们公告的入口页，该入口组件需要加载公告消息列表。Message组件的代码如下：

```
var React = require('react-native');
var Util = require('./util');
var Item = require('./message/item');
var Detail = require('./message/detail');
var Service = require('./service');

var {
  View,
  Text,
  ScrollView,
  StyleSheet,
  TextInput,
  Image,
  TouchableOpacity,
} = React;

var Message = React.createClass({
  render: function(){
    var contents = [];
    var items = [];
    if(this.props.data.status){
      contents = this.props.data.data;
    }
    for(var i = 0; i < contents.length; i++){
      items.push(
        <Item
          data={contents[i]}
          nav={this.props.navigator}
          component={Detail}
          text={contents[i].message}
          name={contents[i].username}
          date={contents[i].time}/>
      );
    }

    return (
      <ScrollView style={styles.container}>
        <View style={{height:50,padding:7,}}>
          <TextInput style={styles.search} placeholder="搜索"/>
        </View>
        <View style={{backgroundColor:'#fff', borderTopWidth:1, borderTopColor:'#ddd'}}>
          {items}
          <View style={{height:35}}></View>
        </View>
      </ScrollView>
    );
  }
```

```
});

var styles = StyleSheet.create({
  container:{
    flex:1,
    backgroundColor:'#F5F5F5',
    flexDirection:'column'
  },
  search:{
    height:35,
    borderWidth:Util.pixel,
    borderColor:'#ccc',
    paddingLeft:10,
    borderRadius:6,
    backgroundColor:'#fff',
  }
});

module.exports = Message;
```

我们在入口组件中使用items数组来添加列表项（即Item组件）。Item组件（address_book/views/message/item.js）的代码如下：

```
var React = require('react-native');
var Util = require('../util');
var Service = require('../service');

var {
  View,
  Text,
  StyleSheet,
  Image,
  TouchableOpacity,
} = React;

var Item = React.createClass({
  render: function(){
    return (
      <TouchableOpacity onPress={this.loadPage.bind(this, this.props.data)}>
        <View style={styles.item}>
          <View style={styles.width55}>
            <Text style={{color:'#fff', fontSize:18,fontWeight:'bold'}}>
              {this.props.name.substr(0,1)}</Text>
          </View>
          <View style={{flexDirection:'column',flex:1}}>
            <Text numberOfLines={2} style={styles.text}>
              {this.props.text}
            </Text>
            <Text style={styles.date}>
              {this.props.date}
            </Text>
          </View>
```

```
            <View numberOfLines={1} style={styles.m10}>
              <Text style={styles.name}>{this.props.name}</Text>
            </View>
          </View>
        </TouchableOpacity>
    );
  },
  loadPage: function(data){
    var content = data;
    this.props.nav.push({
      title:'消息详情',
      component: this.props.component,
      passProps:{
        content: content
      }
    });
  }
});

var styles = StyleSheet.create({
  item:{
    height:80,
    padding:5,
    borderBottomWidth: Util.pixel,
    borderBottomColor: '#ddd',
    flexDirection:'row',
    alignItems:'center',
  },
  img:{
    width:50,
    height:50,
    borderRadius:4,
  },
  width55:{
    width:50,
    height:50,
    borderRadius:4,
    marginLeft:10,
    alignItems:'center',
    justifyContent:'center',
    backgroundColor: '#05C147',
    marginRight:10,
  },
  text:{
    flex:1,
    marginBottom:5,
    opacity:0.7
  },
  date:{
    color:'#ccc',
    fontSize:11,
  },
  m10:{
    marginLeft:10
```

```
    },
    name:{
      color:'#929292',
      fontSize:13
    }
});

module.exports = Item;
```

当点击列表项时，Item组件的_loadPage函数加载公告详情页。公告列表页的效果如图8-16所示。

图8-16　公告消息列表

现在我们已经完成公告消息列表，接下来需要完成公告消息的详情页。address_book/views/message/detail.js就是详情页组件，即Detail，具体代码如下：

```
var React = require('react-native');
var {
  View,
  Text,
  StyleSheet,
  Image,
  ScrollView,
  TouchableOpacity,
} = React;

var Detail = React.createClass({
  render: function(){
```

```
            var content = this.props.content;
            return (
              <ScrollView>

                  <View style={styles.content}>
                      <Text style={{lineHeight:20,}}>{content.message}</Text>
                  </View>

                  <View style={[styles.luokuan, {marginTop:25}]}>
                      <View style={{flex:1}}></View>
                      <Text style={[styles.text, {color:'#007AFF'}]}>{content.username}</Text>
                  </View>

                  <View style={styles.luokuan}>
                      <View style={{flex:1}}></View>
                      <Text style={[styles.text, {color:'#3BC1FF'}]}>{content.time}</Text>
                  </View>

              </ScrollView>
            );
        }
});

var styles = StyleSheet.create({
    content:{
        marginTop:20,
        marginLeft:15,
        marginRight:15,
        opacity:0.85,
    },
    luokuan:{
        flex:1,
        flexDirection:'row',
        marginRight:20,
    },
    text:{
        lineHeight:20,
        width:90
    }
});
module.exports = Detail;
```

详情页的效果如图8-17所示。

图8-17 消息详情页

8.4.7 管理功能

address_book/views/manager.js是管理页面的入口，即Manager组件，其代码如下所示：

```
var React = require('react-native');
var Util = require('./util');
var AddUser = require('./manager/addUser');
var ModifyPassword = require('./manager/modifyPassword');
var DeleteUser = require('./manager/deleteUser');
var PostMessage = require('./manager/postMessage');

var {
  View,
  Text,
  ScrollView,
  StyleSheet,
  TouchableOpacity,
  AsyncStorage,
} = React;

var Manager = React.createClass({
  render: function(){
    var colors = ['#F4000B', '#17B4FF', '#FFD900', '#F00000'];
```

```
        var tags = ['U', 'A', 'D', 'M'];
        var items = ['修改密码', '增加联系人', '删除联系人', '发布公告'];
        var components = [ModifyPassword, AddUser, DeleteUser, PostMessage];
        var JSXDOM = [];
        for(var i in items){
          JSXDOM.push(
            <TouchableOpacity onPress={this._loadPage.bind(this, components[i], items[i])}>
              <View style={[styles.item, {flexDirection:'row'}]}>
                <Text style={[styles.tag, {color: colors[i]}]}>{tags[i]}</Text>
                <Text style={[styles.font,{flex:1}]}>{items[i]}</Text>
              </View>
            </TouchableOpacity>
          );
        }


        return (
          <ScrollView style={styles.container}>
            <View style={styles.wrapper}>
              {JSXDOM}
            </View>

            <View style={{marginTop:30}}>
              <TouchableOpacity onPress={this._clear}>
                <View style={[styles.item, {flexDirection:'row'}]}>
                  <Text style={[styles.tag, {color: colors[i]}]}>Q</Text>
                  <Text style={[styles.font,{flex:1}]}>退出登录</Text>
                </View>
              </TouchableOpacity>
            </View>
          </ScrollView>
        );
    },

    _loadPage: function(component, title){
      this.props.navigator.push({
        title: title,
        component: component
      });
    },

    _clear: function(){
      this.props.navigator.pop();
      AsyncStorage.clear();
    }

});

var styles = StyleSheet.create({
  container:{
    flex:1,
    backgroundColor:'#F5F5F5',
  },
  item:{
```

```
        height:40,
        justifyContent: 'center',
        borderTopWidth: Util.pixel,
        borderTopColor: '#ddd',
        backgroundColor:'#fff',
        alignItems:'center',
    },
    font:{
        fontSize:15,
        marginLeft:5,
        marginRight:10,
    },
    wrapper:{
        marginTop:30,
    },
    tag:{
        marginLeft:10,
        fontSize:16,
        fontWeight:'bold'
    }
});
module.exports = Manager;
```

这里我们将所有组件放在一个数组中，即 components = [ModifyPassword, AddUser, DeleteUser, PostMessage]。然后遍历数组，生成功能列表，完成后的效果如图8-18所示。

图8-18　管理功能列表

下面简要介绍这4个功能。

1. 修改密码功能

这个功能需要用户传入原始密码和新密码，同时需要传入token，验证token的合法性。在address_book/views/manager/modifyPassword.js文件中添加如下代码：

```
var React = require('react-native');
var Util = require('./../util');
var Service = require('../service');

var {
  View,
  Text,
  ScrollView,
  StyleSheet,
  TouchableOpacity,
  TextInput,
  AlertIOS,
  AsyncStorage,
} = React;

var ModifyUser = React.createClass({

  render: function(){
    return (
      <ScrollView>

        <View style={{height:35, marginTop:30,}}>
          <TextInput style={styles.input} password={true} placeholder="原始密码"
            onChangeText={this._getOldPassword}/>
        </View>

        <View style={{height:35,marginTop:5}}>
          <TextInput style={styles.input} password={true} placeholder="新密码"
            onChangeText={this._getNewPassword}/>
        </View>

        <View>
          <TouchableOpacity onPress={this._resetPassword}>
            <View style={styles.btn}>
              <Text style={{color:'#FFF'}}>修改密码</Text>
            </View>
          </TouchableOpacity>
        </View>
      </ScrollView>
    );
  },

  _getOldPassword: function(val){
    this.setState({
      oldPassword: val
    });
  },
```

```
_getNewPassword: function(val){
  this.setState({
    password: val
  });
},

_resetPassword: function(){
  var path = Service.host + Service.updatePassword;
  var that = this;
  //需要服务器端确认登录的token，只有token一致才能有权限修改密码
  AsyncStorage.getItem('token', function(err, data){
    if(!err){
      Util.post(path, {
        password: that.state.password,
        oldPassword: that.state.oldPassword,
        token: data,
      }, function(data){
        if(data.status){
          AlertIOS.alert('成功', data.data);
        }else{
          AlertIOS.alert('失败', data.data);
        }
      });
    }else{
      AlertIOS.alert('失败', data.data);
    }
  });
}

});

var styles = StyleSheet.create({
  input:{
    flex:1,
    marginLeft:20,
    marginRight:20,
    height:35,
    borderWidth:1,
    borderColor:'#ddd',
    borderRadius:4,
    paddingLeft:5,
    fontSize:13,
  },
  btn:{
    justifyContent:'center',
    alignItems:'center',
    marginTop:20,
    backgroundColor:'#1DB8FF',
    height:38,
    marginLeft:20,
    marginRight:20,
    borderRadius:4,
  }
});
```

```
module.exports = ModifyUser;
```

修改密码界面完成的效果如图8-19所示。

图8-19 修改密码

2. 增加用户功能

增加用户功能也是独立组件，即address_book/views/manager/addUser.js，具体代码如下：

```
var React = require('react-native');
var Util = require('./../util');
var Service = require('./../service');

var {
  View,
  Text,
  ScrollView,
  StyleSheet,
  TouchableOpacity,
  TextInput,
  PickerIOS,
  AlertIOS,
} = React;

var AddUser = React.createClass({

  getInitialState: function(){
    var items = ['A', 'B', 'C', 'D', 'E', 'F'];
    var tags = ['框架研发', 'BU产品', 'BU研发', '启明星', '项目管理', '公共产品'];
```

```
        return {
          items: items,
          tags: tags,
          selectA:{
            backgroundColor:'#3BC1FF',
            borderColor:'#3BC1FF'
          },
          select_A:{
            color: '#FFF'
          },
          yan:{
            backgroundColor:'#3BC1FF',
            borderColor:'#3BC1FF'
          },
          yan_text:{
            color: '#FFF'
          },
          tag:'研发',
          partment: '框架研发'
        };
    },
    render: function(){
      var tagOne = [];
      for(var i = 0; i <3; i++){
        tagOne.push(
          <TouchableOpacity onPress={this._select.bind(this, this.state.items[i])}>
            <View style={[styles.part, this.state['select' + this.state.items[i]]]}>
              <Text style={this.state['select_' + this.state.items[i]]}>
                {this.state.tags[i]}</Text>
            </View>
          </TouchableOpacity>
        );
      }

      var tagTwo = [];
      for(var i = 3; i <6; i++){
        tagTwo.push(
          <TouchableOpacity onPress={this._select.bind(this, this.state.items[i])}>
            <View style={[styles.part, this.state['select' + this.state.items[i]]]}>
              <Text style={this.state['select_' + this.state.items[i]]}>
                {this.state.tags[i]}</Text>
            </View>
          </TouchableOpacity>
        );
      }

      return (
        <ScrollView style={{paddingTop:30}}>
          <View style={styles.row}>
            <Text style={styles.label}>用户名</Text>
            <TextInput style={styles.input} onChangeText={this._setUserName}/>
          </View>
```

```
        <View style={styles.row}>
          <Text style={styles.label}>密码</Text>
          <TextInput style={styles.input} password={true}
            placeholder="初始密码" onChangeText={this._setPassword}/>
        </View>

        <View style={styles.row}>
          <Text style={styles.label}>邮箱</Text>
          <TextInput style={styles.input} onChangeText={this._setEmail}/>
        </View>

        <View style={styles.row}>
          <Text style={styles.label}>电话</Text>
          <TextInput style={styles.input} onChangeText={this._setTel}/>
        </View>

        <View style={styles.partment}>
          {tagOne}
        </View>
        <View style={styles.partment}>
          {tagTwo}
        </View>

        <View style={{marginTop:30,flexDirection:'row', justifyContent:'center'}}>
          <TouchableOpacity onPress={this._selectType.bind(this, 'yan')}>
            <View style={[styles.part, this.state.yan]}>
              <Text style={this.state.yan_text}>研发</Text>
            </View>
          </TouchableOpacity>

          <TouchableOpacity onPress={this._selectType.bind(this, 'chan')}>
            <View style={[styles.part, this.state.chan]}>
              <Text style={this.state.chan_text}>产品</Text>
            </View>
          </TouchableOpacity>

          <TouchableOpacity onPress={this._selectType.bind(this, 'project')}>
            <View style={[styles.part, this.state.project]}>
              <Text style={this.state.project_text}>项目</Text>
            </View>
          </TouchableOpacity>
        </View>

        <View style={{marginTop:30, alignItems:'center', justifyContent:'center'}}>
          <TouchableOpacity onPress={this._addUser}>
            <View style={styles.btn}>
              <Text>创建用户</Text>
            </View>
          </TouchableOpacity>
        </View>
      </ScrollView>
    );
  },
  _select: function(id){
```

```javascript
var obj = {};
var color = {};
var items = {
  A:{},
  B:{},
  C:{},
  D:{},
  E:{},
  F:{}
};
//加上选中效果
obj['select' + id] = {
  backgroundColor:'#3BC1FF',
  borderColor:'#3BC1FF'
};
color['select_' + id] = {
  color: '#fff',
};
this.setState(obj);
this.setState(color);
this.setState();
//清除其他选中效果
delete items[id];
for(var i in items){
  var newObj = {};
  newObj['select' + i] = {
    backgroundColor:'#FFF',
    borderColor:'#ddd'
  };
  var newColor = {};
  newColor['select_' + i] = {
    color: '#000',
  };
  this.setState(newObj);
  this.setState(newColor);
}
//增加变量
var partment = '框架研发';
switch (id){
  case 'A':
    partment = this.state.tags[0];
    break;
  case 'B':
    partment = this.state.tags[1];
    break;
  case 'C':
    partment = this.state.tags[2];
    break;
  case 'D':
    partment = this.state.tags[3];
    break;
  case 'E':
    partment = this.state.tags[4];
    break;
```

```js
      case 'F':
        partment = this.state.tags[5];
        break;
    }
    this.setState({
      partment: partment
    });
  },
  _selectType: function(id){
    var obj = {};
    var color = {};
    var items = {
      yan:{},
      chan:{},
      project:{}
    };
    //加上选中效果
    obj[id] = {
      backgroundColor:'#3BC1FF',
      borderColor:'#3BC1FF'
    };
    color[id + '_text'] = {
      color: '#fff',
    };
    this.setState(obj);
    this.setState(color);

    //清除其他选中效果
    delete items[id];
    for(var i in items){
      var newObj = {};
      newObj[i] = {
        backgroundColor:'#FFF',
        borderColor:'#ddd'
      };
      var newColor = {};
      newColor[i + '_text'] = {
        color: '#000',
      };
      this.setState(newObj);
      this.setState(newColor);
    }
    //增加变量
    var tag = '研发';
    switch (id){
      case 'yan':
        tag = '研发';
        break;
      case 'chan':
        tag = '产品';
        break;
      case 'project':
        tag = '项目';
        break;
```

```
        default :
          break;
      }
      this.setState({
        tag: tag
      });
    },
    _setUserName:function(val){
      this.setState({
        username: val
      });
    },
    _setPassword:function(val){
      this.setState({
        password: val
      });
    },
    _setEmail:function(val){
      this.setState({
        email: val
      });
    },
    _setTel: function(val){
      this.setState({
        tel: val
      });
    },
    _addUser: function(){
      var username = this.state.username;
      var email = this.state.email;
      var password = this.state.password;
      var partment = this.state.partment;
      var tag = this.state.tag;
      var tel = this.state.tel;

      if(!username || !email || !password || !tel){
        return AlertIOS.alert('提示', '用户名、初始密码、邮箱电话、必填, 请确认!');
      }
      var obj = {
        username: username,
        email: email,
        password: password,
        partment: partment,
        tag: tag,
        tel: tel
      };
```

```
      var path = Service.host + Service.createUser;
      Util.post(path, obj, function(data){
        if(data.status){
          AlertIOS.alert('成功','创建用户成功，请告知用户初始密码');
        }else{
          AlertIOS.alert('失败','创建用户失败');
        }
      });

    }

});

var styles = StyleSheet.create({
  row:{
    flexDirection:'row',
    alignItems:'center',
    marginBottom:7,
  },
  label:{
    width:50,
    marginLeft:10,
  },
  input:{
    borderWidth: Util.pixel,
    height:35,
    flex:1,
    marginRight:20,
    borderColor:'#ddd',
    borderRadius: 4,
    paddingLeft:5,
    fontSize:14,
  },
  partment:{
    flexDirection:'row',
    justifyContent:'center',
    marginTop:10,
  },
  part:{
    width:65,
    height:30,
    borderWidth:Util.pixel,
    borderColor: '#ddd',
    borderRadius:3,
    alignItems:'center',
    justifyContent:'center',
    marginRight:10
  },
  btn:{
    borderColor:'#268DFF',
    height:35,
    width:200,
    borderRadius:5,
    borderWidth:Util.pixel,
```

```
        alignItems:'center',
        justifyContent:'center'
    }
});
module.exports = AddUser;
```

这里需要注意的是，单选按钮的实现代码比较累赘，可以参考https://github.com/vczero/react-native-tab-menu/blob/master/tab.js进行修改。最终实现的增加用户界面的效果如图8-20所示。

图8-20　增加用户

3. 删除用户

这里我们需要根据用户的邮箱删除用户，具体代码如下：

```
var React = require('react-native');
var Util = require('./../util');
var Service = require('../service');

var {
    View,
    Text,
    ScrollView,
    StyleSheet,
    TouchableOpacity,
    TextInput,
    AlertIOS,
    AsyncStorage,
} = React;
```

```
var DeleteUser = React.createClass({
  render: function(){
    return (
      <ScrollView>

          <View style={{height:35, marginTop:30,}}>
            <TextInput style={styles.input}  placeholder="请输入用户的邮箱"
              onChangeText={this._getEmail}/>
          </View>

          <View>
            <TouchableOpacity onPress={this._deleteUser}>
              <View style={styles.btn}>
                <Text style={{color:'#FFF'}}>删除用户</Text>
              </View>
            </TouchableOpacity>
          </View>
      </ScrollView>
    );
  },

  _getEmail: function(val){
    this.setState({
      email: val
    });
  },

  _deleteUser: function(){
    var that = this;
    AlertIOS.alert('提示', '确认删除该用户？', [
      {text: '删除', onPress: function(){
        var path = Service.host + Service.deleteUser;
        AsyncStorage.getItem('token', function(err, data){
          if(!err){
            Util.post(path,{
              token: data,
              email: that.state.email
            }, function(data){
              if(data.status){
                AlertIOS.alert('成功', '删除成功');
              }else{
                AlertIOS.alert('失败', '删除失败');
              }
            });
          }else{
            AlertIOS.alert('提示', '没有权限');
          }
        });
      }
      },
      {text: '取消', onPress: ()=>null},
    ]);
  }
```

```
});
var styles = StyleSheet.create({
  input:{
    flex:1,
    marginLeft:20,
    marginRight:20,
    height:35,
    borderWidth:1,
    borderColor:'#ddd',
    borderRadius:4,
    paddingLeft:5,
    fontSize:13,
  },
  btn:{
    justifyContent:'center',
    alignItems:'center',
    marginTop:20,
    backgroundColor:'#1DB8FF',
    height:38,
    marginLeft:20,
    marginRight:20,
    borderRadius:4,
  }
});
module.exports = DeleteUser;
```

删除用户界面的最终效果如图8-21所示。

图8-21　删除用户

4. 发布公告

最后调用发布公告接口发布公告，例如通知同事新同学的加入，具体代码如下：

```
var React = require('react-native');
var Service = require('./../service');
var Util = require('./../util');

var {
  View,
  TextInput,
  ScrollView,
  StyleSheet,
  TouchableOpacity,
  Image,
  Text,
  AsyncStorage
} = React;

var PostMessage = React.createClass({

  render: function(){
    return (
      <ScrollView >
        <View>
          <TextInput multiline={true}
                     onChangeText={this._onChange}
                     style={styles.textinput}
                     placeholder="请输入公告内容"/>
        </View>
        <View style={{marginTop:20}}>
          <TouchableOpacity onPress={this._postMessage}>
            <View style={styles.btn}>
              <Text style={{color:'#fff'}}>发布公告</Text>
            </View>
          </TouchableOpacity>
        </View>
      </ScrollView>
    );
  },

  _onChange: function(val){
    if(val){
      this.setState({
        message: val
      });
    }
  },

  _postMessage: function(){
    var that = this;
    AsyncStorage.getItem('token', function(err, token){
      if(err){
```

```
          alert('权限失效,请退出App,重新登录');
        }else{
          Util.post(Service.host + Service.addMessage, {
            token: token,
            message: that.state.message
          }, function(data){
            if(data.status){
              alert('添加成功!');
            }else{
              alert('添加失败!');
            }
          });
        }
    });
  }
});
var styles = StyleSheet.create({
  textinput:{
    flex:1,
    height:100,
    borderWidth:1,
    borderColor:'#ddd',
    marginTop:30,
    marginLeft:20,
    marginRight:20,
    paddingLeft:8,
    fontSize:13,
    borderRadius:4
  },
  btn:{
    flex:1,
    justifyContent:'center',
    alignItems:'center',
    backgroundColor:'#1DB8FF',
    height:38,
    marginLeft:20,
    marginRight:20,
    borderRadius:4,
  }
});
module.exports = PostMessage;
```

发布公告界面完成的效果如图8-22所示。

图8-22 发布公告

8.4.8 关于

关于页面一般是介绍我们的项目，这里我们使用WebView组件跳转到GitHub项目页面，相关文件为address_book/views/about.js，具体代码如下：

```
var React = require('react-native');
var webview = require('./about/webview');

var {
  View,
  Text,
  ScrollView,
  StyleSheet,
  Image,
  TouchableOpacity,
} = React;


var About = React.createClass({

  render: function(){
    return (
      <ScrollView style={styles.container}>

        <View style={styles.wrapper}>
          <Image style={styles.avatar} source={require('image!me_1')}></Image>
```

```
            <Text style={{fontSize:14, marginTop:10, color:'#ABABAB'}}>Author: vczero</Text>
            <Text style={{fontSize:14, marginBottom:20, color:'#ABABAB'}}>
              Version: v0.0.1</Text>

            <View style={{flexDirection:'row'}}>
              <TouchableOpacity onPress=
                {this._openWebView.bind(this, 'https://github.com/vczero/React-Native-App')}>
                <Image style={styles.img} source={require('image!github')}/>
              </TouchableOpacity>

              <TouchableOpacity onPress=
                {this._openWebView.bind(this, 'http://weibo.com/vczero')}>
                <Image style={[styles.img, {width:25,height:25}]}
                  source={require('image!weibo')}/>
              </TouchableOpacity>
            </View>
          </View>

        </ScrollView>
      );
    },

    _openWebView: function(url){
      this.props.navigator.push({
        title:'项目地址',
        component: webview,
        passProps:{
          url: url
        }
      });
    }
});

var styles = StyleSheet.create({
  container:{
    flex:1,
  },
  wrapper:{
    alignItems:'center',
    marginTop:50,
  },
  avatar:{
    width:90,
    height:90,
    borderRadius:45,
  },
  img:{
    width:20,
    height:20,
    marginRight:5,
  }
});

module.exports = About;
```

这里我们调用了WebView组件（详见address_book/views/about/webview/js），具体代码如下：

```
var React = require('react-native');
var {
  WebView,
  ScrollView,
  Text,
  View,
} = React;

var webview = React.createClass({
  render: function(){
    return(
      <View style={{flex:1}}>
        <WebView url={this.props.url}/>
      </View>
    );
  }
});

module.exports = webview;
```

8.4.9 建议

到目前为止，我们算是开发完这个项目了。我们可以在服务器端使用node app.js启动服务器端项目，然后在Xcode中使用快捷键cmd+R运行React Native项目，接着就能看到最终效果了。

"百灵鸟"App虽已完成，但还有很多需要完善的地方。目前"百灵鸟"App托管在GitHub上，地址是https://github.com/vczero/React-Native-App，我们希望大家一起继续完善其功能。

第9章
基于LBS的应用开发

在第8章中，我们不仅开发了React Native客户端，还开发了Node.js服务端，一步步实现了"百灵鸟"这款简单的应用。作为开发者，我们深知技术只有运用于生产才能产生价值，只有通过实战才能将技术掌握得更深。在这一章中，我们将继续实战的风格，开发一款基于LBS的应用——附近。

目前，O2O市场十分火爆，各巨头都在O2O领域深耕布局。比如，"美团外卖"和"饿了么"大肆补贴观众。我们打开"美团外卖"，首先会定位位置，然后显示附近的商家列表，这个功能就是LBS功能。一般情况下，我们不会去构建基础的LBS服务，就像定位服务一样，我们不会为了地址解析去构建庞大的地址库。在实际的开发中，一般会选择类似高德地图、百度地图、Google地图提供的API来实现LBS功能。在这一章中，我们会借助高德地图提供的API来实现"附近"应用。

9.1 功能设计

在开发App之前，我们都需要确定需求。而需求确认中最重要的环节是功能设计和细节把控。对于我们个人开发程序而言，灵活度要高得多，但是在开发App之前都会思考，这款App应该包含哪些功能。我们不需要去做PRD（产品需求文档）评审，因为很多功能和技术实现都已经在脑海中进行了认证。我们需要做的就是勾勒有哪些功能以及相关功能需要展示的信息。

9.1.1 需求确定

"附近"应用应该有什么需求呢？这里我们选择几个经常性的需求——卫生间、银行、餐饮和电影院。比如，我们到了一个地方，想要看看附近的取款机在哪，需要知道离自己最近的卫生间在哪，需要看看附近有没有什么吃饭的地方等。这里我们开发的"附近"应用需要的功能如下。

- ❏ 显示用户附近的公厕列表（含道路、距离、所属单位等信息）。
- ❏ 显示用户附近的银行列表（含ATM、银行名称、距离、道路信息，如果有电话，则显示电话）。
- 显示用户附近的餐饮列表（含商家名称、餐饮类型、电话、道路信息，如果有电话，则显示电话）。
- 显示用户附近的电影院列表（含电影院名称、距离、道路等信息）。
- 列表的显示条数控制在10条以内。
- 所有列表都有基本的详情页。
- 显示的电话号码能够拨号。
- 所有列表项都在对应的地图上展示。

为了方便开发，需要画出简单的原型，如图9-1所示。

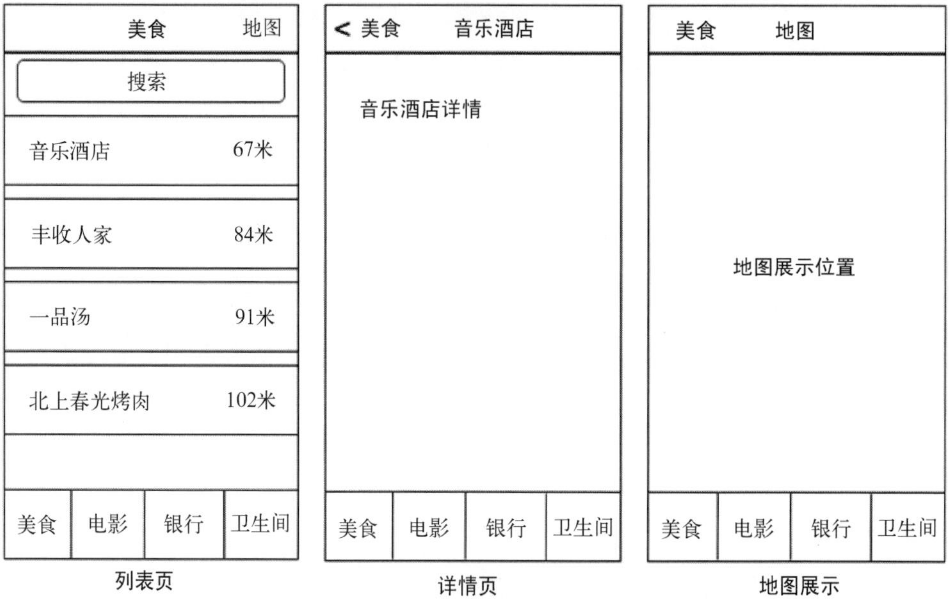

图9-1 原型

9.1.2 开发目录结构

针对原型，我们设计的代码结构如图9-2所示。

9.2 程序入口和工具模块

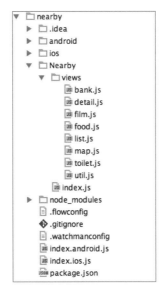

图9-2 代码结构

在上述结构中，index.ios.js是入口文件，Nearby是附近功能的代码目录。Nearby/index.js是附近功能的入口文件，Nearby/views/util.js是工具模块，Nearby/views/map.js是WebView地图模块，Nearby/views/list.js是列表组件，Nearby/views/detail.js是详情页组件，Nearby/views/food.js是餐饮组件，Nearby/views/film.js是电影组件，Nearby/views/toilet.js是卫生间组件，Nearby/views/bank.js是银行组件。

9.2 程序入口和工具模块

开发React Native程序的第一步是编写入口文件，即index.ios.js代码。在入口文件中，我们需要加载入口组件，同时还需要开发好工具类作为公共模块调用。

9.2.1 注册应用程序

在index.ios.js中，我们加载Nearby的入口模块。这里我们使用AppRegistry.registerComponent注册应用程序，具体代码如下：

```
'use strict';
var React = require('react-native');
var Nearby = require('./Nearby/index');

var {
  AppRegistry
} = React;
AppRegistry.registerComponent('nearby', () => Nearby);
```

9.2.2 工具模块

为了方便开发，我们需要开发一个工具模块，该工具模块包含单位像素、POST方法、GET方法、高德地图API key等，具体代码如下：

```javascript
var React = require('react-native');
var Dimensions = require('Dimensions');

var {
  PixelRatio
} = React;

var Util = {
  //单位像素
  pixel: 1 / PixelRatio.get(),
  //屏幕尺寸
  size: {
    width: Dimensions.get('window').width,
    height: Dimensions.get('window').height
  },

  //POST方法
  post: function (url, data, callback) {
    var fetchOptions = {
      method: 'POST',
      headers: {
        'Accept': 'application/json',
        'Content-Type': 'application/json'
      },
      body: JSON.stringify(data)
    };

    fetch(url, fetchOptions)
      .then((response) => response.text())
      .then((responseText) => {
        callback(JSON.parse(responseText));
      });
  },

  //GET方法
  getJSON: function(url, callback){
    fetch(url)
      .then((response) => response.text())
      .then((responseText) => {
        callback(JSON.parse(responseText));
      });
  },
  //高德地图key，测试key，请勿商用
  amapKey: '98cd4d3c1c2865132e73d851654c9c1b',
  //周边搜索服务
  searchURL: 'http://restapi.amap.com/v3/place/around?',
```

```
    detailURL: 'http://restapi.amap.com/v3/place/detail?'
};
module.exports = Util;
```

因为这里选择的是高德地图API，所以需要申请开发者key。在上面的代码中，amapKey是作为测试使用，后期可能会变动。因此，我们需要按照以下步骤完成key的申请。首先，可以到http://id.amap.com/member/注册开发者账号。注册成功后，打开http://lbs.amap.com/console/控制台，选择获取key。这里，我们需要申请两个key：一个是Web服务API的key，它主要用于调用附近的搜索服务和详情服务；另一个是JavaScript API的key，用于在WebView里面使用HTML+JS调用高德地图。申请key的选择界面如图9-3所示。

图9-3　申请 key

9.2.3　Nearby 组件入口

Nearby/index.js是"附近"的功能组件入口，我们需要在该组件中加载其他组件的路由以及定义一些全局变量，具体代码如下：

```
var React = require('react-native');
var Bank = require('./views/bank');
var Film = require('./views/film');
var Food = require('./views/food');
var Toilet = require('./views/toilet');
var Map = require('./views/map');

var {
  AppRegistry,
  StyleSheet,
  Text,
  View,
  ScrollView,
  NavigatorIOS,
  StatusBarIOS,
  TabBarIOS
} = React;
```

```
//是否开启真实的定位？如果开启了_GEO_OPEN，则_GEO_TEST_POS会失效
_GEO_OPEN = true;
//模拟定位数据
_GEO_TEST_POS = '121.390686,31.213976';

//高亮
StatusBarIOS.setStyle('light-content');
//开启网络状态
StatusBarIOS.setNetworkActivityIndicatorVisible(true);

var Nearby = React.createClass({
  getInitialState: function(){
    return{
      selected: '美食'
    };
  },
  render: function() {
    return (
      <View style={styles.container}>
        <TabBarIOS>
          <TabBarIOS.Item
            title='美食'
            selected={this.state.selected === '美食'}
            icon={require("image!food")}
            onPress={()=>{this.setState({selected: '美食'})}}>
            <NavigatorIOS
              barTintColor='#007AFF'
              titleTextColor="#fff"
              tintColor="#fff"
              ref="nav_food"
              style={styles.container}
              initialRoute={{
                component: Food,
                title: '美食',
                rightButtonTitle: '地图',
                onRightButtonPress: ()=>{
                  this.refs.nav_food.navigator.push({
                    title: '地图',
                    component: Map,
                    passProps:{
                      type:'餐饮'
                    }
                  });
                }
              }}
            />
          </TabBarIOS.Item>

          <TabBarIOS.Item
            title='电影'
            selected={this.state.selected === '电影'}
            icon={require("image!film")}
            onPress={()=>{this.setState({selected: '电影'})}}>
            <NavigatorIOS
```

```
        style={styles.container}
        barTintColor='#007AFF'
        titleTextColor="#fff"
        tintColor="#fff"
        ref="nav_film"
        initialRoute={{
          component: Film,
          title: '电影',
          rightButtonTitle: '地图',
          onRightButtonPress: ()=>{
            this.refs.nav_film.navigator.push({
              title: '地图',
              component: Map,
              passProps:{
                type:'电影院'
              }
            });
          }
        }}
      />
</TabBarIOS.Item>

<TabBarIOS.Item
  title='银行'
  selected={this.state.selected === '银行'}
  icon={require("image!bank")}
  onPress={()=>{this.setState({selected: '银行'})}}>
    <NavigatorIOS
      style={styles.container}
      barTintColor='#007AFF'
      titleTextColor="#fff"
      tintColor="#fff"
      ref="nav_bank"
      initialRoute={{
        component: Bank,
        title: '银行',
        rightButtonTitle: '地图',
        onRightButtonPress: ()=>{
          this.refs.nav_bank.navigator.push({
            title: '地图',
            component: Map,
            passProps:{
              type:'银行'
            }
          });
        }

      }}
    />
</TabBarIOS.Item>

<TabBarIOS.Item
  title='卫生间'
  selected={this.state.selected === '卫生间'}
```

```jsx
              icon={require("image!toilet")}
              onPress={()=>{this.setState({selected: '卫生间'})}}>
              <NavigatorIOS
                style={styles.container}
                barTintColor='#007AFF'
                titleTextColor="#fff"
                tintColor="#fff"
                ref="nav_toilet"
                initialRoute={{
                  component: Toilet,
                  title: '卫生间',
                  rightButtonTitle: '地图',
                  onRightButtonPress: ()=>{
                    this.refs.nav_toilet.navigator.push({
                      title: '地图',
                      component: Map,
                      passProps:{
                        type:'厕所'
                      }
                    });
                  }
                }}
              />
            </TabBarIOS.Item>
          </TabBarIOS>
        </View>
      );
    },
  });

  var styles = StyleSheet.create({
    container: {
      flex: 1
    }
  });
  module.exports = Nearby;
```

这里我们定义了_GEO_OPEN,表示是否开启真实定位。因为模拟器无法真实地模拟用户位置,所以在模拟器中调试时使用_GEO_TEST_POS = '121.390686,31.213976'模拟定位。我们使用StatusBarIOS.setStyle('light-content')将App的状态栏设置为白色,这样在导航栏为深色时,能够高亮显示手机的状态。StatusBarIOS.setNetworkActivityIndicatorVisible(true)表示开启手机网络状态,如果网络状态不佳,则会出现loading的效果。此外,我们会使用TabBarIOS.Item表现每一个tab项。TabBarIOS.Item使用的图标可以到https://github.com/vczero/React-Native-App/tree/master/pic/nearby下载,然后在Xcode中将bank.png、film.png、food.png、toilet.png拖曳进Images.xcassets中,如图9-4所示。

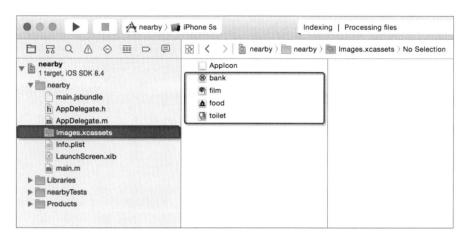

图9-4　图标资源

我们在getInitialState中默认加载"美食"Tab项，即将this.state.selected的默认值设置为"美食"。当用户触发TabBarIOS.Item的onPress事件时，使用this.setState({selected: '当前Tab的值'})}设置为当前点击Tab的标题。通过selected={this.state.selected === '当前Tab的值'}可以判断该Tab是否被选中。每一个TabBarIOS.Item加载一个NavigatorIOS组件，这样每一个Tab都是单独的路由。在NavigatorIOS组件上设置导航栏的背景颜色为蓝色、字体为白色。同时，我们使用了ref。当点击初始化后的导航栏的右侧"地图"字样时，使用this.refs.ref名称.navigator.push来加载地图WebView，展示列表中每一项在地图上的位置。

9.3　列表组件开发

在App中，列表组件十分常见。因为我们最多显示10条数据，所以开发简单的列表组件即可。如果需要使用ListView组件，可以参考10.4节。

9.3.1　通用列表组件开发

现在来开发列表组件，我们希望列表组件能够承载更多的功能。4个功能（餐饮、电影、银行、卫生间）列表都基于该组件。我们在Nearby/views/list.js中添加如下代码：

```
var React = require('react-native');
var Geolocation = require('Geolocation');
var Util = require('./util');
var Detail = require('./detail');

var {
  View,
  ScrollView,
  Text,
```

```
    StyleSheet,
    TextInput,
    ActivityIndicatorIOS,
    TouchableOpacity,
    LinkingIOS,
    ActionSheetIOS,
    WebView,
    AsyncStorage
} = React;

var List = React.createClass({
    getInitialState: function() {
        return {
            list: null,
            count: 0,
            keywords: ''
        };
    },
    render: function(){
        var items = [];
        if(this.state.list){
            var len = this.state.list.length > 10 ? 10 : this.state.list.length;
            for(var i = 0; i < len; i++){
                var obj = this.state.list[i];
                items.push(
                    <TouchableOpacity style={styles.item}
                        onPress={this._loadDetail.bind(this, obj.id, obj.name)}>
                        <View style={styles.row}>
                            <View style={{flex:1}}>
                                <Text numberOfLines={1} style={styles.name}>{obj.name}</Text>
                                <Text numberOfLines={1} style={styles.type}>{obj.type}</Text>
                            </View>
                            <View style={styles.distance}>
                                <Text numberOfLines={1} style={[styles.mi, {color:'#4C4C4C'}]}>
                                    {obj.distance}米
                                </Text>
                                <Text numberOfLines={1} style={styles.address}>{obj.address}</Text>
                            </View>
                        </View>
                        {
                            obj.tel.length ?
                            (<TouchableOpacity style={styles.phone}
                                onPress={this._call.bind(this, obj.tel)}>
                                <Text numberOfLines={1} >电话</Text>
                            </TouchableOpacity>)
                            :null
                        }
                    </TouchableOpacity>
                );
            }
        }
        var placeholder = '搜索' + this.props.type;
        return (
```

```jsx
        <ScrollView style={styles.container}>
          <View style={styles.searchBg}>
            <TextInput style={styles.input} placeholder={placeholder}
              onChangeText={this._onChangeText}
              onEndEditing={this._onEndEditing}/>
            <View>
              <Text style={styles.tip}>
                已为您筛选
                <Text style={{color:'#FA2530'}}>{this.state.count}</Text>
                条数据
              </Text>
            </View>
          </View>
          {items}
          {items.length? null : <View style={styles.activity}>
            <ActivityIndicatorIOS color="#248BFD"/></View>}
          <View style={{height:40}}></View>
        </ScrollView>
      );
    },
    componentDidMount: function(){
      var that = this;
      Geolocation.getCurrentPosition(function(data){
        var lnglat = data.coords.longitude + ',' + data.coords.latitude;
        AsyncStorage.setItem('pos', lnglat);
        var url = Util.searchURL + 'key=' + Util.amapKey + '&keywords='
          + that.props.type + '&extensions=base';
        if(_GEO_OPEN){
          url += '&location=' + lnglat;
          that._doGetData(url);
        }else{
          url += '&location=' + _GEO_TEST_POS;
          that._doGetData(url);
        }
      }, function(err){
        alert('定位失败，请重新开启应用定位');
      });
    },

    _doGetData: function(url){
      var that = this;
      Util.getJSON(url, function(data){
        if(data.status && data.info === 'OK'){
          var count = data.pois.length > 10? 10: data.pois.length;
          that._addStorage(data);
          that.setState({
            list: data.pois,
            count: count
          });
        }else{
          alert('没有查询到相应的数据');
        }
      });
    },
```

```
/*加载详情页*/
_loadDetail: function(id, name){
  this.props.nav.push({
    component: Detail,
    title: name,
    passProps:{
      id: id
    }
  });
},
_onChangeText: function(val){
  this.setState({
    keywords: val
  });
},
_onEndEditing: function(){
  var that = this;
  var keywords = this.state.keywords;
  var url = Util.searchURL + 'key=' + Util.amapKey + '&keywords='
    + keywords + '&types=' + that.props.type + '&extensions=base';
  that.setState({
    list: null
  });
  AsyncStorage.getItem('pos', function(err, result){
    if(_GEO_OPEN){
      if(!err){
        url += '&location=' + result;
        that._doGetData(url);
      }else{
        alert('定位失败');
      }
    }else{
      url += '&location=' + _GEO_TEST_POS;
      that._doGetData(url);
    }
  });
},

//添加到本地存储
_addStorage: function(data){
  var posArr = [];
  var len = data.pois.length > 10? 10: data.pois.length;
  for(var i = 0; i < len; i++){
    posArr.push(data.pois[i].location);
  }
  var posStr = posArr.join(',');
  AsyncStorage.setItem('_' + this.props.type , posStr);
},

//拨打电话
_call: function(tel){
  if(tel.length){
```

```
        var arr = tel.split(';');
        var BUTTONS = [];
        for(var i in arr){
          BUTTONS.push(arr[i]);
        }
        BUTTONS.push('取消');

        ActionSheetIOS.showActionSheetWithOptions({
          options: BUTTONS,
          cancelButtonIndex: BUTTONS.length - 1
        }, function(index){
          arr[index] && LinkingIOS.openURL('tel://' + arr[index]);
        });
      }else{
        alert('没有提供号码');
      }
    }

});

var styles = StyleSheet.create({
  container:{
    flex:1,
    backgroundColor:'#ddd'
  },
  input:{
    height:38,
    marginLeft:10,
    marginRight:10,
    borderWidth:Util.pixel,
    paddingLeft:5,
    marginTop:10,
    borderColor: '#868687',
    borderRadius:3,
    fontSize:15
  },
  tip:{
    fontSize:12,
    marginLeft:10,
    marginTop:5,
    color: '#505050'
  },
  row:{
    flexDirection:'row',
    marginLeft:10,
    marginRight:10,
    marginTop:10,
    paddingTop:5
  },
  distance:{
    width:120,
    alignItems:'flex-end',
  },
  name:{
```

```
      fontSize:15,
      marginBottom:6
    },
    type:{
      fontSize:12,
      color:'#686868'
    },
    mi:{
      fontSize:12,
      color:'#686868'
    },
    address:{
      fontSize:12,
      marginTop:5,
      color:'#686868'
    },
    phone:{
      marginLeft:10,
      marginRight:10,
      height:30,
      marginTop:10,
      justifyContent:'center',
      alignItems:'center',
      borderWidth:Util.pixel,
      borderColor:'#ccc',
      borderRadius:2,
    },
    searchBg:{
      backgroundColor:'#fff',
      paddingBottom:10
    },
    item:{
      marginTop:10,
      backgroundColor:'#fff',
      paddingBottom:10,
      borderTopWidth:Util.pixel,
      borderBottomWidth:Util.pixel,
      borderColor:'#ccc'
    },
    activity:{
      marginTop:50,
      justifyContent:'center',
      alignItems:'center',
    }
});

module.exports = List;
```

这里我们加载的是Geolocation API，这样就可以获取到用户的位置。然后在componentDidMount中获取用户位置并存储到AsyncStorage中：

```
componentDidMount: function(){
  var that = this;
```

```
Geolocation.getCurrentPosition(function(data){
  var lnglat = data.coords.longitude + ',' + data.coords.latitude;
  AsyncStorage.setItem('pos', lnglat);
  var url = Util.searchURL + 'key=' + Util.amapKey + '&keywords='
    + that.props.type + '&extensions=base';
  if(_GEO_OPEN){
    url += '&location=' + lnglat;
    that._doGetData(url);
  }else{
    url += '&location=' + _GEO_TEST_POS;
    that._doGetData(url);
  }
}, function(err){
  alert('定位失败，请重新开启应用定位');
});
}
```

我们可以看到这里使用了Geolocation.getCurrentPosition来获取用户当前的位置信息。我们将位置信息拼接成服务请求的URL串。如果是模拟器环境，则模拟定位；如果是真机环境，则使用真实的定位信息。我们将url传递给_doGetData函数来获取服务返回的数据，其代码如下：

```
_doGetData: function(url){
  var that = this;
  Util.getJSON(url, function(data){
    if(data.status && data.info === 'OK'){
      var count = data.pois.length > 10? 10: data.pois.length;
      that._addStorage(data);
      that.setState({
        list: data.pois,
        count: count
      });
    }else{
      alert('没有查询到相应的数据');
    }
  });
}
```

在上述代码中，_doGetData调用了我们在Util中封装的getJSON方法，返回最多10条数据。我们使用_addStorage函数将返回的数据添加到AsyncStorage中。同时设置显示的列表数据为data.pois（兴趣点），以及设置了多少条数据结果。之所以将返回的数据添加到AsyncStorage中，是为了在地图显示之前方便取到列表的数据集。

同时，我们封装了_call方法，使用LinkingIOS.openURL('tel:// 电话号码')来呼起系统的拨号功能。

为了获得更好的用户体验，我们开发了搜索功能。在用户结束输入框输入时（即onEnd-Editing）发起请求，更新数据存储和视图渲染：

```
_onEndEditing: function(){
  var that = this;
```

```
    var keywords = this.state.keywords;
    var url = Util.searchURL + 'key=' + Util.amapKey + '&keywords='
      + keywords + '&types=' + that.props.type + '&extensions=base';
    that.setState({
      list: null
    });
    AsyncStorage.getItem('pos', function(err, result){
      if(_GEO_OPEN){
        if(!err){
          url += '&location=' + result;
          that._doGetData(url);
        }else{
          alert('定位失败');
        }
      }else{
        url += '&location=' + _GEO_TEST_POS;
        that._doGetData(url);
      }
    });
}
```

上述代码使用了AsyncStorage来获取用户当前的位置信息，这样就不用重复定位了。在_onEndEditing方法中，将拼接的新服务URL串传递给_doGetData方法，从而更新存储和视图。

我们通过传递nav属性获取navigator对象，这样就可以加载详情页，具体代码如_loadDetail方法所示：

```
/*加载详情页*/
_loadDetail: function(id, name){
  this.props.nav.push({
    component: Detail,
    title: name,
    passProps:{
      id: id
    }
  });
}
```

在_loadDetail方法中，id表示POI ID，即点击列表中某一项的ID，这样我们就可以在列表页通过ID来查询该兴趣点的详情。

9.3.2 完成列表页

目前，我们已经完成了通用列表组件。现在需要将通用列表组件应用到"美食"、"电影"、"银行"和"卫生间"这4个功能列表中。

1. "美食"模块

我们加载列表组件，即require('./list')，然后创建Food组件。Food组件主要是传递type和navigator。这样，当加载Food模块时，就会请求附近的"餐饮"信息。"美食"模块的代码在

Nearby/views/food.js文件中，其内容如下：

```
var React = require('react-native');
var List = require('./list');
var Food = React.createClass({
  render: function(){
    return (
      <List type="餐饮" nav={this.props.navigator}/>
    );
  }
});
module.exports = Food;
```

"美食"列表的效果如图9-5所示。

图9-5 "美食"列表

2. "电影"模块

同样，Nearby/views/film.js是电影模块，其代码如下：

```
var React = require('react-native');
var List = require('./list');
var Film = React.createClass({
  render: function(){
    return (
      <List type="电影院" nav={this.props.navigator}/>
    );
  }
});
module.exports = Film;
```

"电影"列表的效果如图9-6所示。

3. "银行"模块

同样，Nearby/views/bank.js是银行模块，其代码如下：

```
var React = require('react-native');
var List = require('./list');
var Bank = React.createClass({
  render: function(){
    return (
      <List type="银行" nav={this.props.navigator}/>
    );
  }
});
module.exports = Bank;
```

图9-6 "电影"列表

"银行"列表的效果如图9-7所示。

4. "卫生间"模块

同样，Nearby/views/toilet.js是卫生间模块，其代码如下：

```
var React = require('react-native');
```

```
var List = require('./list');
var Toilet = React.createClass({
  render: function(){
    return(
      <List type="厕所" nav={this.props.navigator}/>
    );
  }
});
module.exports = Toilet;
```

"卫生间"列表的效果如图9-8所示。

图9-7 "银行"列表

图9-8 "卫生间"列表

9.4 详情页组件开发

详情页组件比较简单,这里只是简单展示了一些详情信息,相关代码在Nearby/views/detail.js文件中,其内容如下:

```
var React = require('react-native');
var Util = require('./util');

var {
  View,
  ScrollView,
  Text,
  StyleSheet,
  TextInput,
  ActivityIndicatorIOS,
  TouchableOpacity
} = React;
```

9.4 详情页组件开发

```
var FoodDetail = React.createClass({
  getInitialState: function(){
    return{
      data: null
    };
  },
  render: function(){
    return (
      <ScrollView>
        {this.state.data?
          <View style={styles.content}>
            <Text style={styles.name}>{this.state.data.name}</Text>
            <Text style={styles.types}>
              类型:
              {this.state.data.type}
            </Text>
            <Text style={styles.address}>
              地址:
              {this.state.data.address}
            </Text>
            <Text style={styles.tag}>
              标签:
              {this.state.data.tag}
            </Text>
            <Text style={styles.server}>
              服务:
              {this.state.data.server}
            </Text>
          </View>
          :null}
      </ScrollView>
    );
  },
  componentDidMount: function(){
    var that = this;
    var url = Util.detailURL + 'key=' + Util.amapKey + '&id='
      + this.props.id + '&extensions=all';
    Util.getJSON(url, function(data){
      if(data.status && data.info === 'OK' && data.pois.length){
        var obj = data.pois[0];
        if(obj.deep_info && obj.deep_info.tag){
          obj.server = obj.deep_info.tag;
        }
        that.setState({
          data: obj
        });
      }else{
        alert('数据服务出错');
      }

    });
  }
});

var styles = StyleSheet.create({
```

```
    container:{
      flex:1
    },
    name:{
      fontSize:15,
      color:'#1D92F5',
      fontWeight:'bold'
    },
    content:{
      marginLeft:10,
      marginRight:10,
      marginTop:10
    },
    tag:{
      fontSize:13,
      marginTop:10
    },
    types:{
      marginTop:10,
      fontSize:13,
      color:'#4C4C4C'
    },
    address:{
      fontSize:13,
      color:'#4C4C4C'
    },
    server:{
      marginTop:10,
      fontSize:13
    }
});
module.exports = FoodDetail;
```

图9-9　详情页

在上述代码中，我们在componentDidMount中获取传递过来的id，然后通过id来拼接请求的URL串，最后通过使用该URL串来完成数据请求。在请求完成的成功回调函数中，使用that.setState来改变数据模型，而数据模型的改变触发视图的重新渲染。详情页的效果如图9-9所示。

9.5　WebView 地图模块开发

我们已经完成了列表页和详情页开发，现在需要将结果项展示在地图上，这样用户就能清楚地知道每一个POI（兴趣点）的位置。这里我们使用高德地图JavaScript API开发HTML5页面，然后将HTML5页面嵌入React Native WebView中。我们需要和WebView协商一个共同的数据传输协议，这里我们使用URL参数传递参数。我们协定的规则是：

```
http://xxx/index.html?pos=121.390686,31.213976&markers=121.390686,31.213976,121.390586,31.213976,
121.350686,31.213976
```

在WebView中加载地图页面时，传递参数pos和markers，其中pos表示当前用户的位置，markers表示地图上的点标记。Nearby/views/map.js就是地图模块，具体代码如下：

```js
var React = require('react-native');
var {
  View,
  Text,
  WebView,
  AsyncStorage
} = React;

var Map = React.createClass({
  getInitialState: function(){
    return{
      url: null
    };
  },
  render: function(){
    var webView = null;
    if(this.state.url){
      webView = <WebView url={this.state.url}/>
    }
    return(
      <View style={{flex:1}}>
        {webView}
      </View>
    );
  },
  componentDidMount: function(){
    var that = this;
    AsyncStorage.multiGet(['_' + that.props.type, 'pos'], function(err, result){
      if(!err){
        var pos = result[1][1];
        var markers = result[0][1];
        var url = 'http://vczero.github.io/webview/index.html?';
        if(_GEO_OPEN){
          url += 'pos=' + pos + '&markers=' + markers;
        }else{
          url += 'pos=' + _GEO_TEST_POS + '&markers=' + markers;
        }
        that.setState({
          url: url
        });
      }else{
        alert('定位失败');
      }
    });
  }
});
module.exports = Map;
```

这里我们使用AsyncStorage.multiGet获取多个key值,然后生成符合规则的URL,加载HTML地图页面。

之所以在9.2.2节中申请高德地图JavaScript API,是因为需要在HTML页面中开发地图应用。这里我们使用申请的key加载JavaScript API:

```
<script type="text/javascript" src="http://webapi.amap.com/maps?v=1.3
    &key=6a1180467b5f36714645d22044535ab7"></script>
<script type="text/javascript">
```

并且通过获取URL参数来设置地图上的Marker（地理标注，地图开发的专业术语），最后根据所有Marker的位置使地图自适应窗口，完整代码如下：

```
<!DOCTYPE html>
<html style="width: 100%;height:100%;">
<head>
    <meta charset="utf-8">
    <meta name="renderer" content="webkit">
    <meta http-equiv="X-UA-Compatible" content="IE=edge,chrome=1">
    <meta name="apple-mobile-web-app-capable" content="yes">
    <meta name="apple-mobile-web-app-status-bar-style" content="black">
    <meta http-equiv="Content-Type" content="text/html; charset=utf-8">
    <meta name="apple-mobile-web-app-title" content="地图">
    <meta name="viewport" content="width=device-width,initial-scale=1,
      maximum-scale=1,minimum-scale=1,minimal-ui">
    <meta name="msapplication-tap-highlight" content="no">
    <title>地图</title>
</head>
<body style="width: 100%;height:100%;padding:0;margin:0;">
    <div id="map" style="width: 100%;height:100%;"></div>
    <script type="text/javascript" src="http://webapi.amap.com/maps?v=1.3
        &key=6a1180467b5f36714645d22044535ab7"></script>
    <script type="text/javascript">
    var url = window.location.href;
    var map = new AMap.Map('map',{
      resizeEnable: true
    });
    map.setZoom(16);
    var urlParams = url.split('?');
    if(urlParams.length > 1){
      var arr = urlParams[1].split('&');
      var obj = {};
      for(var i in arr){
        var kv = arr[i].split('=');
        obj[kv[0]] = kv[1];
      }

      //地图居中
      //index.html?pos=121.390686,31.213976
      if(obj["pos"]){
        var pos = obj["pos"].split(',');
        map.setCenter(new AMap.LngLat(pos[0], pos[1]));
      }
      //添加Marker
      //index.html?pos=121.390686,31.213976&markers=121.390686,
      //31.213976,121.390586,31.213976,121.350686,31.213976
      if(obj["markers"]){
        var marks = obj["markers"].split(',');
        var mks = [];
        for(var i = 0; i < marks.length; i++){
```

```
            if(i % 2 === 0){
                var marker = new AMap.Marker({
                    map: map,
                    position: new AMap.LngLat(marks[i], marks[i + 1])
                });
                mks.push(marker);
            }
        }
        //根据点标记自适应
        map.setFitView(mks);
    }
}
</script>
</body>
</html>
```

地图模块的效果如图9-10所示。

9.6 综合效果

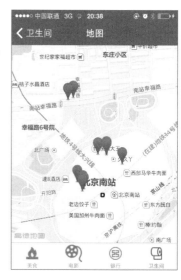

图9-10 地图模块

现在我们已经完成了"附近"这个App的所有功能，其代码已经提交到了GitHub：https://github.com/vczero/React-Native-App，供大家参考。图9-11以"电影院"为例，展示了列表页、详情页和地图。

图9-11 电影院

第10章
豆搜App

第8章介绍了使用React Native和Node.js开发简单的通讯录应用——百灵鸟，第9章介绍了基于高德地图API开发的"附近"App。在这一章中，我们将介绍通过调用豆瓣开放API开发一款图书、电影和音乐搜索App——豆搜。本书所有实例的图片均由豆瓣API提供，版权归豆瓣API所有。

10.1 豆瓣 API

互联网时代是服务共享的时代，调用开放平台的数据服务，不仅可以整合大平台的资源，还可以提高开发效率，也可以节省服务器资源。因此，借用平台的优势，可以帮助我们更加快速地拿到资源，获取用户。比如，开发一款小型的公众App，使用新浪微博的登录接口，不仅可以让用户觉得安全放心，同时也增加了用户量。因为这样可以避免用户重复注册账户，节省用户时间。在这一章中，我们使用的是豆瓣开放API。因为豆瓣在音乐、电影、图书方面的数据相对较全，所以做一款简单的App已经足够了。

10.1.1 熟悉豆瓣 API

一般情况下，开放API的调用是需要申请的，比如新浪微博、开源中国、高德地图开放API、百度地图API等。这里我们使用的API是搜索和详情服务，所以不涉及提交和修改等操作，因此豆瓣也没有针对这些API做权限限制。在"豆搜"App中，一共用到了4个API服务，如下所示：

- 图书搜索
- 图书详情
- 电影搜索
- 音乐搜索

我们可以在豆瓣API官网上（http://developers.douban.com/wiki/?title=api_v2）看到，目前API的版本是V2.0。豆瓣建议大家切换到2.0版本，旧的API不再更新，只会维护。在网站的底部可以看到，豆瓣开放API V2.0提供了不少服务，如图10-1所示。

图10-1　豆瓣API

10.1.2　图书、电影、音乐 API

因为在"豆搜"应用中，我们只用到了图书、电影和音乐相关的服务，所以这里只需要关注前三项即可。

1. 图书API

点击"图书 Api V2"，跳转到图书 API 服务列表（http://developers.douban.com/wiki/?title=book_v2）。这里我们使用的图书服务是图书搜索（即图书API服务列表界面中的"搜索图书"）和图书详情（即图书API服务列表界面中的"获取图书信息"）。

❑ **图书搜索服务接口**。下面是图书搜索服务的URL地址：

```
请求方式：GET
服务地址：https://api.douban.com/v2/book/search
```

该服务的参数如下。

q：表示查询关键字，q和tag必传其一。
tag：查询的标签。
start：取结果的offset（即偏移量），默认为0。
count：取结果的条数，默认为20，最大为100。

例如，我们可以通过浏览器访问https://api.douban.com/v2/book/search?count=10&q=C语言来获取关于C语言查询图书的数据。

- **图书详情服务接口**。下面是图书详情服务的URL地址：

```
请求方式：GET
服务地址：https://api.douban.com/v2/book/:id
```

该服务的参数是id，即图书的编号。

例如，我们可以访问https://api.douban.com/v2/book/1139336获取图书的详细信息。

2. 电影API

点击"电影Api V2"，跳转到电影API服务列表（http://developers.douban.com/wiki/?title=movie_v2）。这里，我们使用的服务是电影搜索，该服务的接口如下：

```
请求方式：GET
服务地址：https://api.douban.com/v2/movie/search
```

该服务的参数如下。

- **q**：表示查询关键字，q和tag必传其一。
- **tag**：查询的标签。
- **start**：取结果的offset（即偏移量），默认为0。
- **count**：取结果的条数，默认为20。

例如，我们可以访问https://api.douban.com/v2/movie/search?count=5&q=幸福来获取关于电影关键字中含"幸福"的数据。

3. 音乐API

点击"音乐Api V2"，跳转到音乐API服务列表（http://developers.douban.com/wiki/?title=music_v2）。这里我们使用的是音乐搜索API，该服务接口如下：

```
请求方式：GET
服务地址：https://api.douban.com/v2/music/search
```

该服务的参数如下。

- **q**：表示查询关键字，q和tag必传其一。
- **tag**：查询的标签。
- **start**：取结果的offset（即偏移量），默认为0。
- **count**：取结果的条数。

例如，我们可以访问https://api.douban.com/v2/music/search?count=10&q=刘德华来获取关于音乐中含"刘德华"的数据。

10.2 应用设计

在开发一款App前,我们都需要设计和架构。比如,要开发一个登录中心,就需要考虑登录的客户端有哪些类型、通用的登录方式是否行得通、各客户端的安全要素是什么等问题。这些问题需要我们在开发前就设计好。如果没有设计和架构,那么在多人协作开发的情况下,很可能出现沟通复杂、代码重复等问题。因此,开发一款App前,做一个简单的架构和设计是必要的。这里,"豆搜"App会从功能设计和模块划分这两个方面来介绍。

10.2.1 功能设计

"豆搜"App是基于豆瓣开放API开发的,功能相对比较简单,主要功能如图10-2所示。

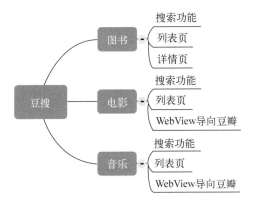

图10-2　功能模块

该App主要包含3个主体功能:图书、电影和音乐。这3个主体功能都提供搜索,方便用户检索需要的内容。我们需要将搜索的结果展示出来,因此需要列表页功能。同时,在一般情况下,我们还需要详情页。比如,要想买一本React Native方面的图书,我们希望能够了解更多的信息,因此,"图书"模块提供了详情页的功能。同时在"电影"和"音乐"模块,我们提供了WebView来展示电影和音乐的详情。

10.2.2 模块划分

10.2.1节已经完成了基本功能的勾勒,现在需要搭建项目。一般而言,为了方便代码管理和维护,通常的做法是将同类业务或者说功能放在同一个文件夹中。这样,模块划分比较明确,后期修改也较为方便。我们搭建的项目结构如图10-3所示。

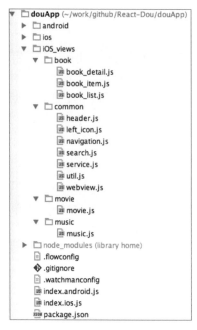

图10-3 项目结构

该项目是基于React Native iOS开发的，因此index.ios.js是入口文件。iOS_views是功能模块的主体文件夹，common是公共组件和功能的文件夹，book是图书功能的文件夹，movie和music分别是电影和音乐功能模块的文件夹。其中每个文件的功能如下。

- book_detail.js：图书详情页组件，它调用了book_items.js模块。
- book_items.js：图书列表项，供图书列表组件和图书详情页组件调用。
- book_list.js：图书列表组件，它调用了search.js模块和header.js模块。
- header.js：公共导航头，供列表组件调用。
- left_icon.js：公共导航头中的回退图标。
- navigation.js：封装Navigator，暴露passProps属性。
- search.js：搜索框组件。
- service.js：服务URL模块。
- util.js：工具类模块，主要包含屏幕尺寸、公共loading组件、GET请求等属性和方法。
- webview.js：该组件在React Native WebView组件的基础上做了进一步封装，添加了头部。这样用户可以使用头部回退按钮，从WebView回退到Native App。
- movie.js：电影列表组件。
- music.js：音乐列表组件。

10.3 公共模块开发

在前两节中,我们已经做了功能的勾勒和模块的划分,现在只需要按照既定的方案进行开发即可。首先,需要完成公共模块的开发。因为工具类和公共模块是业务开发的基础。公共模块对应图10-3中的common文件夹。因为工具类模块是一个文件,这里我们将其放进了common文件夹,作为公共模块的一部分。

10.3.1 工具类开发

在工具类中,我们希望有4个功能和属性是可以暴露给业务开发使用的。很多时候,我们需要根据设备的像素密度获取最小线宽,此时可以使用PixelRatio.get()。同时,在有些情况下,我们需要拿到当前屏幕的宽高,此时可以使用Dimensions.get('window')。相关代码在common/util.js文件中,具体如下:

```
/*!
 *
 * Util模块工具类
 * 主要提供工具方法
 *
 */
var React = require('react-native');
var Dimensions = require('Dimensions');

var {
  PixelRatio,
  ActivityIndicatorIOS
} = React;

module.exports = {
  /*最小线宽*/
  pixel: 1 / PixelRatio.get(),

  /*屏幕尺寸*/
  size: {
    width: Dimensions.get('window').width,
    height: Dimensions.get('window').height
  },
  /**
   * 基于fetch的get方法
   * @method post
   * @param {string} url
   * @param {function} callback 请求成功回调
   */
  get: function(url, successCallback, failCallback){
    fetch(url)
      .then((response) => response.text())
      .then((responseText) => {
        successCallback(JSON.parse(responseText));
```

```
      })
      .catch(function(err){
        failCallback(err);
      });
    },
    /*loading效果*/
    loading: <ActivityIndicatorIOS color="#3E00FF"
      style={{marginTop:40,marginLeft:Dimensions.get('window').width/2-10}}/>
};
```

这里我们在fetch的基础上做了一个简单的GET方法的封装。该方法需要传递三个参数，第一个是请求的服务地址，第二个参数是成功的回调函数，第三个是失败的回调函数。如果请求成功，则将获取的数据转为JSON格式并传递给成功回调函数；如果请求失败，则将错误信息传给失败的回调函数。在util.js模块中，还提供了一个简单的loading组件，方便展示loading效果。

10.3.2 服务列表

"豆搜"App调用的是豆瓣的开放API。为了方便管理这些服务接口，我们将它们放置在一个文件中，即common/service.js，具体的内容如下：

```
/*!
 *
 * 服务URL
 * 基于豆瓣开放API的图书、音乐和电影服务
 * 如果https://api.douban.com/v2/都保持不变，则可以将其设置为BaseURL
 */
module.exports = {
  //图书搜索
  book_search: 'https://api.douban.com/v2/book/search',
  //图书详情
  book_search_id: 'https://api.douban.com/v2/book/',
  //音乐搜索
  music_search: 'https://api.douban.com/v2/music/search',
  //音乐详情
  music_search_id: 'https://api.douban.com/v2/music/',
  //电影搜索
  movie_search: 'https://api.douban.com/v2/movie/search',
  //电影详情
  movie_search_id: 'https://api.douban.com/v2/movie/subject/'
};
```

这样我们在开发过程中就不要过多关心服务地址，只需要service.book_search的形式调用即可。以后，如果服务地址改变，也只用在该文件中修改即可。其实如果https://api.douban.com/v2/保持不变，我们可以定义BaseURL，即var BaseURL = 'https://api.douban.com/v2/';，然后在服务地址中拼接即可。

10.3.3 Navigator 封装

React Native提供了两个导航栏组件，一个是NavigatorIOS，一个是Navigator。NavigatorIOS的封装程度更高，提供了基础的API和属性，开发者简单定制就可以开发一个功能丰富的导航栏。而Navigator相对比较简陋，正是这种简陋，才给了开发者足够的空间去定制。所以说，Navigator的定制能力要高于NavigatorIOS。如果是复杂的导航栏，我们完全可以采用Navigator。在第8章和第9章中，都是使用NavigatorIOS作为导航栏组件，在这一章中，我们将使用Navigator定制简单的导航栏组件。

这里我们封装了简单的common/navigation.js组件，具体代码如下：

```javascript
/*!
 *
 * 封装Navigator
 * 所有的切换过场动画都是从底部往上，回退是从上往下
 * 这里需要注意的是使用{...route.passProps}模仿NavigatorIOS的passProps
 */
var React = require('react-native');
var {
  StyleSheet,
  Text,
  View,
  Navigator
} = React;

module.exports = React.createClass({
  render: function(){
    return (
      <Navigator
        initialRoute={{name: '', component: this.props.component, index:0}}
        configureScene={()=>{return Navigator.SceneConfigs.FloatFromBottom;}}
        renderScene={(route, navigator) => {
          const Component = route.component;
          return (
            <View style={{flex: 1}}>
              <Component navigator={navigator} route={route} {...route.passProps}/>
            </View>
          );
        }}/>
    );
  }
});
```

在上述代码中，我们在initialRoute中传递了component，即this.props.component，这样初始化Navigation组件时，即可传递入口组件。同时，我们使用了{...route.passProps}。这样做的好处有两个：第一个就是可以使用passProps属性，保持和NavigatorIOS一样的接口；第二个就是可以扩展更多的属性。

10.3.4 公共头封装

在一款App中，头部通常是必不可少的，这里我们使用的是定制的头部，其效果如图10-4所示。

图10-4 头部效果

该头部组件可以分为两个部分，一个回退按钮，一个是标题。首先，开发回退按钮组件，即common/left_icon.js，具体代码如下：

```
var React = require('react-native');
var Util = require('./util');

var {
  StyleSheet,
  Text,
  View
} = React;

module.exports = React.createClass({
  render: function(){
    return (
      <View>
        <View style={styles.go}>
        </View>
      </View>
    );
  }
});

var styles = StyleSheet.create({
  go:{
    borderLeftWidth: 4 * Util.pixel,
    borderBottomWidth: 4 * Util.pixel,
    width:15,
    height:15,
    transform: [{rotate: '45deg'}],
    borderColor:'#FFF',
    marginLeft:10
  }
});
```

这里我们使用了transform: [{rotate: '45deg'}]将一个矩形框旋转了45度，并且设置左边框和底边框的宽度为4个像素。这样，在蓝色为底色的头部中，就会显示一个<图标。

接着完成头部组件的主体部分，即common/header.js。在header.js模块中，我们会调用回退按钮组件，具体代码如下：

```
var React = require('react-native');
var Icon = require('./left_icon');
var Util = require('./../../common/util');

var {
  StyleSheet,
  Text,
  View,
  ListView,
  Image,
  ScrollView,
  TouchableOpacity
} = React;

module.exports = React.createClass({
  render: function(){
    var obj = this.props.initObj;
    return (
      <View style={[styles.header, styles.row, styles.center]}>
        <TouchableOpacity style={[styles.row,styles.center]} onPress={this._pop}>
          <Icon/>
          <Text style={styles.fontFFF}>{obj.backName}</Text>
        </TouchableOpacity>
        <View style={[styles.title, styles.center]}>
          <Text style={[styles.fontFFF, styles.titlePos]}
            numberOfLines={1}>{obj.title}</Text>
        </View>
      </View>
    );
  },

  _pop: function(){
    this.props.navigator.pop();
  }
});

var styles = StyleSheet.create({
  row:{
    flexDirection:'row'
  },
  header:{
    height:50,
    backgroundColor:'#3497FF'
  },
  fontFFF:{
    color:'#fff',
    fontSize:17,
    fontWeight:'bold'
  },
  title:{
    flex:1
  },
  titlePos:{
    marginLeft:-20,
```

```
            width:200
    },
    center:{
        justifyContent:'center',
        alignItems:'center'
    }
});
```

头部组件需要传递两个参数——backName和title,其中backName是回退按钮的一部分(即回退的标题),title是该页面的标题。这里我们使用TouchableOpacity包裹了回退按钮和回退标题。当回退部分被点击时,绑定了点击事件(即onPress={this._pop}),这样我们就可以通过this.props.navigator.pop();回退到前一个页面。

10.3.5　WebView封装

这里我们使用WebView嵌入了豆瓣的WebApp页面,同时会将该页面作为全页面展示。为了更好地控制WebView和原生App之间的跳转,我们将WebView统一包装,提供给开发者头部和全页面的WebView。我们希望调用封装好的WebView组件后的效果如图10-5所示。

图10-5　WebView效果

我们封装的WebView组件的代码在common/webview.js文件中,具体如下:

```
var React = require('react-native');
var Util = require('./util');
```

```
var Header = require('./header');

var {
  WebView,
  View
} = React;
module.exports = React.createClass({
  render: function(){
    return (
      <View>
        <Header
          navigator={this.props.navigator}
          initObj={{
            backName: this.props.backName,
            title: this.props.title
          }}/>
        <WebView
          contentInset={{top:-40}}
          startInLoadingState={true}
          style={{width: Util.size.width, height:Util.size.height -50}}
          url={this.props.url}></WebView>
      </View>
    );
  }
});
```

这里我们调用了工具类（common/util.js）和公共头组件（common/header.js）。工具类模块主要提供屏幕大小，size属性用于计算WebView页面的大小。头部组件提供回退到App的功能。在WebView中，我们使用了contentInset和startInLoadingState属性。contentInset属性主要用于WebView内嵌页面的偏移，这里向上偏移40，是为了隐藏豆瓣WebApp的头部，防止出现两个头部（即我们定义的公共头组件和豆瓣WebApp头部）。这里将startInLoadingState设置为true，这样在加载页面的时候会出现loading效果，避免App白屏，减少用户的疑虑。

10.3.6 搜索框封装

因为搜索框在三个列表页中都会用到，所以需要开发通用的搜索框组件。但是，我们也不希望过度封装，这样就削弱了开发者定制的能力。这里仅仅指定搜索框的高度、边框颜色、边框线宽等，输入框的事件全部交由调用者控制。搜索框组件即（common/search.js）的代码如下：

```
var React = require('react-native');
var Util = require('./util');
var {
  StyleSheet,
  Text,
  View,
  TabBarIOS,
  TextInput
} = React;
```

```
module.exports = React.createClass({
  render: function(){
    return (
      <View style={styles.flex_1}>
        <TextInput style={[styles.flex_1, styles.input]} {...this.props}/>
      </View>
    );
  }
});
var styles = StyleSheet.create({
  flex_1:{
    flex:1
  },
  input:{
    borderWidth:Util.pixel,
    height:40,
    borderColor:'#DDDDDD',
    paddingLeft:5
  }
});
```

这里我们使用了{...this.props}，这样调用者在调用搜索框组件时挂载的属性都可以传递给TextInput。

10.4 功能开发

公共组件已经开发完成，现在就开始开发App的功能了。因为图书的功能模块相对较多，所以这里重点介绍一下。而电影和音乐模块的详情页都交由WebView内嵌豆瓣WebApp了，所以相对简单。

10.4.1 入口组件

"豆搜" App是基于React Native iOS开发的，所以index.ios.js是入口组件。入口组件一般需要做的事情是：初始化路由和整个App布局。我们在入口组件中使用了TabBarIOS组件和封装的Navigation组件，这样TabBarIOS可以用于不同功能模块的切换，Navigation可以作为路由的入口。入口组件的代码如下：

```
var React = require('react-native');
var Navigation = require('./iOS_views/common/navigation');
var Book = require('./iOS_views/book/book_list');
var Music = require('./iOS_views/music/music');
var Movie = require('./iOS_views/movie/movie');

var {
  AppRegistry,
```

```
    StyleSheet,
    Text,
    View,
    TabBarIOS,
    ScrollView,
    StatusBarIOS
} = React;

StatusBarIOS.setHidden(true);
var douApp = React.createClass({
  getInitialState: function(){
    return {
      selectedTab: '图书'
    };
  },
  render: function() {
    return (
      <TabBarIOS>
        <TabBarIOS.Item
          title="图书"
          selected={this.state.selectedTab === '图书'}
          icon={require('image!book')}
          onPress={() => {
            this.setState({
              selectedTab: '图书'
            });
          }}>
          <Navigation component={Book}/>
        </TabBarIOS.Item>

        <TabBarIOS.Item
          title="电影"
          selected={this.state.selectedTab === '电影'}
          icon={require('image!movie')}
          onPress={() => {
            this.setState({
              selectedTab: '电影'
            });
          }}>
          <Navigation component={Movie}/>
        </TabBarIOS.Item>

        <TabBarIOS.Item
          title="音乐"
          selected={this.state.selectedTab === '音乐'}
          icon={require('image!music')}
          onPress={() => {
            this.setState({
              selectedTab: '音乐'
            });
          }}>
          <Navigation component={Music}/>
        </TabBarIOS.Item>
      </TabBarIOS>
```

```
    );
  }
});

AppRegistry.registerComponent('douApp', () => douApp);
```

在上述代码中,我们通过 StatusBarIOS.setHidden(true);隐藏了状态栏,同时使用 this.state.selectedTab 来切换选项卡。这里,在每一个 TabBarIOS.Item 中都使用了 Navigation 组件。所以,Navigation 的初始路由都会在 TabBarIOS.Item 切换时展现。TabBarIOS 的效果如图 10-6 所示。

图 10-6 TabBarIOS 效果

10.4.2 图书列表页开发

图书列表页包含两部分,一个是搜索功能,另一个是列表功能。但是搜索功能其实和列表功能是有联系的,因为通过关键字搜索出来的结果需要在列表中展示。因此,我们需要完成的第一步是列表项组件,即 book/book_items.js。

1. 列表项组件

因为列表项组件同样会提供给图书详情页使用,所以封装列表项组件时要灵活些,不能在列表项组件内部绑定点击事件,否则在详情页就会出现循环引用的问题。列表项组件的代码如下:

```
var React = require('react-native');
var Util = require('./../common/util');
var {
  StyleSheet,
  Text,
  View,
  ListView,
  Image,
  TouchableOpacity
} = React;

module.exports = React.createClass({
  render: function(){
    var row = this.props.row;
    return(
      <TouchableOpacity style={[styles.row, styles.item]} {...this.props}>
        <View style={[styles.center]}>
          <Image source={{uri: row.image}} style={styles.book_img}/>
        </View>
        <View style={styles.content}>
          <View>
            <Text style={{width:200}} numberOfLines={1}>{row.title}</Text>
```

```
          </View>
          <View style={{marginTop:10}}>
            <Text style={[styles.publisher, {width:200}]}
              numberOfLines={1}>{row.publisher}</Text>
          </View>
          <View style={{marginTop:10}}>
            <Text style={[styles.publisher, {width:200}]}
              numberOfLines={1}>{row.author}</Text>
          </View>
          <View style={[styles.row,{marginTop:10}]}>
            <Text style={styles.price}>{row.price}</Text>
            <Text style={styles.pages}>{row.pages}页</Text>
          </View>
        </View>
      </TouchableOpacity>
    );
  }
});

var styles = StyleSheet.create({
  row:{
    flexDirection:'row'
  },
  item:{
    height:120,
    borderTopWidth:Util.pixel,
    borderBottomWidth:Util.pixel,
    marginTop:5,
    marginBottom:5,
    borderColor:'#ccc'
  },
  book_img:{
    width:80,
    height:100,
    resizeMode:Image.resizeMode.contain
  },
  center:{
    justifyContent:'center',
    alignItems:'center'
  },
  content:{
    marginTop:10,
    marginLeft:10
  },
  publisher:{
    color:'#A3A3A3',
    fontSize:13
  },
  price:{
    color:'#2BB2A3',
    fontSize:16
  },
  pages:{
    marginLeft:10,
```

```
      color:'#A7A0A0'
    }
});
```

列表项组件主要负责渲染图书的简要信息，比如图书的图片、书名、出版社、作者、价格和页码。这里也使用了{...this.props}，这样调用者就可以自己决定是否需要点击事件。列表项组件完成的效果如图10-7所示。

图10-7　列表项组件

2. 搜索和列表开发

开发完了列表项组件，开发列表就是水到渠成的事了。我们希望列表页的头部是搜索框，搜索框下面是列表。列表组件（即book/book_list.js）的代码如下：

```
var React = require('react-native');
var Search = require('./../common/search');
var Util = require('./../common/util');
var ServiceURL = require('./../common/service');
var BookItem = require('./book_item');
var BookDetail = require('./book_detail');

var {
  StyleSheet,
  Text,
  View,
  ListView,
  Image,
  ScrollView,
  ActivityIndicatorIOS,
  TouchableOpacity
} = React;

module.exports = React.createClass({
  getInitialState: function() {
    var ds = new ListView.DataSource({rowHasChanged: (r1, r2) => r1 !== r2});
    return {
      dataSource: ds.cloneWithRows([]),
      keywords: 'c语言',
      show: false
    };
  },
  render: function(){
    return(
      <ScrollView style={styles.flex_1}>

        <View style={[styles.search, styles.row]}>
```

```jsx
            <View style={styles.flex_1}>
              <Search placeholder="请输入图书的名称" onChangeText={this._changeText}/>
            </View>
            <TouchableOpacity style={styles.btn} onPress={this._search}>
              <Text style={styles.fontFFF}>搜索</Text>
            </TouchableOpacity>
          </View>
          {
            this.state.show ?
            <ListView
              dataSource={this.state.dataSource}
              renderRow={this._renderRow}
            />
            : Util.loading
          }
        </ScrollView>
      );
    },
    componentDidMount: function(){
      this.getData();
    },
    //渲染图书列表项
    _renderRow: function(row){
      return (
        <BookItem row={row} onPress={this._loadPage.bind(this, row.id)}/>
      );
    },
    _changeText: function(val){
      this.setState({
        keywords: val
      });
    },
    _search: function(){
      this.getData();
    },
    //根据关键字查询
    getData: function(){
      var ds = new ListView.DataSource({rowHasChanged: (r1, r2) => r1 !== r2});
      var that = this;
      var baseURL = ServiceURL.book_search + '?count=10&q=' + this.state.keywords;
      //开启loading
      this.setState({
        show: false
      });
      Util.get(baseURL, function(data){
        if(!data.books || !data.books.length){
          return alert('图书服务出错');
        }
        var books = data.books;
        that.setState({
          dataSource: ds.cloneWithRows(books),
          show: true
        });
      }, function(err){
        alert(err);
```

```
      });
    },
    _loadPage: function(id){
      this.props.navigator.push({
        component: BookDetail,
        passProps:{
          id: id
        }
      });
    }
});
var styles = StyleSheet.create({
  flex_1:{
    flex:1,
    marginTop:5
  },
  search:{
    paddingLeft:5,
    paddingRight:5,
    height:45
  },
  btn:{
    width:50,
    backgroundColor:'#0091FF',
    justifyContent:'center',
    alignItems:'center'
  },
  fontFFF:{
    color:'#fff'
  },
  row:{
    flexDirection:'row'
  }
});
```

在上述代码中，我们在getInitialState方法中返回了dataSource、keywords和show属性，其中dataSource是渲染列表的数据源，keywords表示搜索的关键字，show表示是否显示列表。

在render方法中，我们将搜索功能包裹在View中，并且为Search组件绑定onChangeText事件，也为"搜索"按钮绑定了onPress事件。this.state.show控制是显示列表还是显示loading。在ListView组件中，dataSource是ListView组件的数据源，renderRow方法用于渲染列表的每一项，即渲染book/book_items.js组件。

在componentDidMount中，我们调用了getData方法。getData方法的作用是请求图书搜索服务，然后将获取的数据设置到dataSource并且关闭loading。在搜索功能模块，绑定的onPress事件即this._search调用的也是getData方法，以便更新数据源。

在_renderRow方法中，我们渲染了BookItem组件，并且绑定了点击事件，传递图书的id。_loadPage是根据图书id加载图书的详情页。这里之所以能够使用this.props.navigator.push传递passProps属性，是因为10.3.3节做了Navigator的封装，即挂载了route的passProps属性。图书列表页开发完成的效果如图10-8所示。

图10-8　图书列表页

10.4.3　图书详情页开发

现在，我们需要开发图书的详情页，即book/book_detail.js模块。在详情页，需要做的就是拼装头部组件和列表项组件，添加图书简介和作者简介，具体代码如下：

```
var React = require('react-native');
var Util = require('./../common/util');
var ServiceURL = require('./../common/service');
var BookItem = require('./book_item');
var Header = require('./../common/header');

var {
  StyleSheet,
  Text,
  View,
  ListView,
  Image,
  ScrollView,
  TouchableOpacity
} = React;

module.exports = React.createClass({
  getInitialState: function(){
```

```
        return{
          data: null
        };
     },
     render: function(){
       return(
         <ScrollView style={styles.m10}>
           {
             this.state.data ?
               <View>
                 <Header
                   navigator={this.props.navigator}
                   initObj={{
                      backName: '图书',
                      title: this.state.data.title
                   }}/>
                 <BookItem row={this.state.data}/>
                 <View>
                   <Text style={[styles.title]}>图书简介</Text>
                   <Text style={styles.text}>{this.state.data.summary}</Text>
                 </View>

                 <View>
                   <Text style={[styles.title]}>作者简介</Text>
                   <Text style={styles.text}>{this.state.data.author_intro}</Text>
                 </View>
                 <View style={{height:50}}></View>
               </View>
               : Util.loading
           }
         </ScrollView>
       );
     },

     componentDidMount: function(){
       var id = this.props.id;
       var that = this;
       var url = ServiceURL.book_search_id + '/' + id;
       Util.get(url, function(data){
         that.setState({
           data: data
         });
       }, function(err){
         alert(err);
       });
     }
});

var styles = StyleSheet.create({
  m10:{
    flex:1
  },
  title:{
    fontSize:16,
    marginLeft:10,
```

```
        marginTop:10,
        marginBottom:10
    },
    text:{
        marginLeft:10,
        marginRight:10,
        color:'#000D22'
    }
});
```

这里我们使用了Header组件，用于回退和标题的显示。同时加载了BookItem组件来显示图书的基本信息。同时，this.state.data.summary和this.state.data.author_intro分别用于显示图书简介和作者信息。我们在componentDidMount中请求了图书详情服务，即var url = ServiceURL.book_search_id + '/' + id;，然后将数据传递给state来渲染详情页视图。图书详情页的效果如图10-9所示。

图10-9　图书详情页

10.4.4　电影模块开发

开发完图书模块，电影模块的开发就要简单很多。同样，电影模块分为搜索、列表和详情页这3个功能。但是电影的详情页是WebView内嵌豆瓣WebApp来实现的。电影模块的实现放在movie/movie.js文件中，其内容如下：

```
var React = require('react-native');
```

```javascript
var Search = require('./../common/search');
var Util = require('./../common/util');
var ServiceURL = require('./../common/service');
var webView = require('./../common/webview');

var {
  StyleSheet,
  Text,
  View,
  ListView,
  Image,
  ScrollView,
  ActivityIndicatorIOS,
  TouchableOpacity
} = React;

module.exports = React.createClass({
  getInitialState: function() {
    var ds = new ListView.DataSource({rowHasChanged: (r1, r2) => r1 !== r2});
    return {
      dataSource: ds.cloneWithRows([]),
      keywords: '幸福',
      show: false
    };
  },
  render: function(){
    return(
      <ScrollView style={styles.flex_1}>

        <View style={[styles.search, styles.row]}>
          <View style={styles.flex_1}>
            <Search placeholder="请输入电影名称" onChangeText={this._changeText}/>
          </View>
          <TouchableOpacity style={styles.btn} onPress={this._search}>
            <Text style={styles.fontFFF}>搜索</Text>
          </TouchableOpacity>
        </View>
        {
          this.state.show ?
            <ListView
              dataSource={this.state.dataSource}
              renderRow={this._renderRow}
            />
            : Util.loading
        }

      </ScrollView>
    );
  },

  componentDidMount: function(){
    this._getData();
  },

  _changeText: function(val){
```

```
    this.setState({
      keywords: val
    });
},
_search: function(){
    this._getData();
},
_renderRow: function(row){
    var casts = row.casts;
    var names = [];
    for(var i in casts){
      names.push(casts[i].name);
    }

    return (
      <View style={[styles.row,styles.item]}>
        <View>
          <Image style={styles.img} source={{uri: row.images.medium}}/>
        </View>
        <View>
          <Text style={styles.textWitdh} numberOfLines={1}>
            名称：{row.title}
          </Text>
          <Text style={styles.textWitdh} numberOfLines={1}>
            演员：{names}
          </Text>
          <Text style={styles.textWitdh} numberOfLines={1}>
            评分：{row.rating.average}
          </Text>
          <Text style={styles.textWitdh} numberOfLines={1}>
            时间：{row.year}
          </Text>
          <Text style={styles.textWitdh} numberOfLines={1}>
            标签：{row.genres}
          </Text>
          <TouchableOpacity style={styles.goDou}
            onPress={this._goDouBan.bind(this, row.title, row.alt)}>
            <Text>详情</Text>
          </TouchableOpacity>
        </View>
      </View>
    );
},

_getData: function(){
    var ds = new ListView.DataSource({rowHasChanged: (r1, r2) => r1 !== r2});
    var that = this;
    var baseURL = ServiceURL.movie_search + '?count=10&q=' + this.state.keywords;
    this.setState({
      show: false
    });
    Util.get(baseURL, function(data){
      if(!data.subjects || !data.subjects.length){
```

```
          return alert('电影服务出错');
        }
        var subjects = data.subjects;
        that.setState({
          dataSource: ds.cloneWithRows(subjects),
          show: true
        });
      }, function(err){
        alert(err);
      });
    },

    _goDouBan: function(title, url){
      this.props.navigator.push({
        component: webView,
        passProps:{
          backName: '电影',
          title: title,
          url: url
        }
      });
    }

  });

  var styles = StyleSheet.create({
    flex_1:{
      flex:1,
      marginTop:5
    },
    search:{
      paddingLeft:5,
      paddingRight:5,
      height:45
    },
    btn:{
      width:50,
      backgroundColor:'#0091FF',
      justifyContent:'center',
      alignItems:'center'
    },
    fontFFF:{
      color:'#fff'
    },
    row:{
      flexDirection:'row'
    },
    img:{
      width:80,
      height:110,
      resizeMode: Image.resizeMode.contain
    },
    textWitdh:{
      width:200,
      marginLeft:10
```

```
        },
        item:{
            marginTop:10,
            height:140,
            paddingTop:15,
            paddingLeft:10,
            borderBottomWidth:Util.pixel,
            borderTopWidth:Util.pixel,
            borderColor:"#ddd"
        },
        goDou:{
            justifyContent:'center',
            alignItems:'center',
            height:32,
            width:60,
            borderWidth:Util.pixel,
            borderColor:'#3C9BFD',
            marginLeft:30,
            marginTop:10,
            borderRadius:3
        }
});
```

同样，我们在电影的列表页中使用了公共的搜索框组件，调用_getData方法将获取的电影信息展示在列表中。ListView组件的renderRow方法用于渲染列表的每一项。Text组件的numberOfLines属性表示显示多少行，如果给定了Text组件的宽度，那么多余的文本将以省略号...表示。列表页完成的效果如图10-10所示。

图10-10　电影列表页

这里同样需要详情页，我们在"详情"按钮上绑定了onPress事件，即this._goDouBan.bind(this, row.title, row.alt)。在_goDouBan方法中，我们将路由导向了封装的WebView组件，并且传递了回退按钮中的文字、头部的标题以及打开页面的标题。详情页的效果如图10-11所示。

图10-11　电影详情页

10.4.5　音乐模块开发

电影模块和音乐模块（music/music.js）十分类似，除了列表项的渲染外。音乐模块的代码如下：

```
var React = require('react-native');
var Search = require('./../common/search');
var Util = require('./../common/util');
var ServiceURL = require('./../common/service');
var webView = require('./../common/webview');

var {
  StyleSheet,
  Text,
  View,
  ListView,
  Image,
  ScrollView,
  ActivityIndicatorIOS,
  TouchableOpacity
} = React;
```

```jsx
module.exports = React.createClass({
  getInitialState: function() {
    var ds = new ListView.DataSource({rowHasChanged: (r1, r2) => r1 !== r2});
    return {
      dataSource: ds.cloneWithRows([]),
      keywords: '偏偏喜欢你',
      show: false
    };
  },
  render: function(){
    return(
      <ScrollView style={styles.flex_1}>

        <View style={[styles.search, styles.row]}>
          <View style={styles.flex_1}>
            <Search placeholder="请输入歌曲/歌手名称" onChangeText={this._changeText}/>
          </View>
          <TouchableOpacity style={styles.btn} onPress={this._search}>
            <Text style={styles.fontFFF}>搜索</Text>
          </TouchableOpacity>
        </View>
        {
          this.state.show ?
            <ListView
              dataSource={this.state.dataSource}
              renderRow={this._renderRow}
            />
          : Util.loading
        }

      </ScrollView>
    );
  },

  componentDidMount: function(){
    this._getData();
  },

  _changeText: function(val){
    this.setState({
      keywords: val
    });
  },

  _search: function(){
    this._getData();
  },

  _renderRow: function(row){
    return (
      <View style={styles.item}>
        <View style={styles.center}>
          <Image style={styles.img} source={{uri: row.image}}/>
        </View>
        <View style={[styles.row]}>
          <Text style={[styles.flex_1,{marginLeft:20}]}>
```

```
            numberOfLines={1}>曲目：{row.title}</Text>
          <Text style={[styles.textWidth]} numberOfLines={1}>演唱：{row.author}</Text>
        </View>
        <View style={[styles.row]}>
          <Text style={[styles.flex_1, {marginLeft:20}]}
            numberOfLines={1}>时间：{row.attrs['pubdate']}</Text>
          <Text style={styles.textWidth} numberOfLines={1}>评分：
            {row['rating']['average']}</Text>
        </View>
        <View style={[styles.center]}>
          <TouchableOpacity style={[styles.goDou, styles.center]}
            onPress={this._goDouBan.bind(this, row.title, row.mobile_link)}>
            <Text>详情</Text>
          </TouchableOpacity>
        </View>
      </View>
    );
  },

  _getData: function(){
    var ds = new ListView.DataSource({rowHasChanged: (r1, r2) => r1 !== r2});
    var that = this;
    var baseURL = ServiceURL.music_search + '?count=10&q=' + this.state.keywords;
    this.setState({
      show: false
    });
    Util.get(baseURL, function(data){
      if(!data.musics || !data.musics.length){
        return alert('音乐服务出错');
      }
      var musics = data.musics;
      that.setState({
        dataSource: ds.cloneWithRows(musics),
        show: true
      });
    }, function(err){
      alert(err);
    });
  },

  _goDouBan: function(title, url){
    this.props.navigator.push({
      component: webView,
      passProps:{
        title: title,
        url: url,
        backName: '音乐'
      }
    });
  }
});

var styles = StyleSheet.create({
  flex_1:{
    flex:1,
    marginTop:5
  },
```

```
    search:{
      paddingLeft:5,
      paddingRight:5,
      height:45
    },
    btn:{
      width:50,
      backgroundColor:'#0091FF',
      justifyContent:'center',
      alignItems:'center'
    },
    fontFFF:{
      color:'#fff'
    },
    row:{
      flexDirection:'row'
    },
    img:{
      width:70,
      height:70,
      borderRadius:35
    },
    center:{
      justifyContent:'center',
      alignItems:'center'
    },
    item:{
      marginTop:10,
      borderTopWidth:Util.pixel,
      borderBottomWidth:Util.pixel,
      borderColor:'#ddd',
      paddingTop:10,
      paddingBottom:10
    },
    textWidth:{
      width:120
    },
    goDou:{
      height:35,
      width:60,
      borderWidth:Util.pixel,
      borderColor:'#3082FF',
      borderRadius:3
    }
});
```

这里我们同样使用了搜索组件和WebView组件。在_renderRow方法中，我们渲染了音乐的名称、歌手、出版时间、评分等。列表的效果如图10-12所示。

在_goDouBan方法中，我们将封装的WebView组件推送进路由。该WebView内嵌的是音乐的详情页，具体效果如图10-13所示。

图10-12　音乐列表　　　　　　图10-13　音乐详情页

10.5　完成豆搜 App

至此，豆搜App的开发已经完成。虽然这是一个简单的App，但是很好地体现了React Native组件化的开发思想。本章是全书的最后一章，也是终结篇。学习到此，应该可以给自己开发一款有趣的React Native App了。伙伴们，一起努力前行吧，加油！

欢迎加入 图灵社区 iTuring.cn

——最前沿的IT类电子书发售平台

电子出版的时代已经来临。在许多出版界同行还在犹豫彷徨的时候,图灵社区已经采取实际行动拥抱这个出版业巨变。作为国内第一家发售电子图书的IT类出版商,图灵社区目前为读者提供两种DRM-free的阅读体验:在线阅读和PDF。

相比纸质书,电子书具有许多明显的优势。它不仅发布快,更新容易,而且尽可能采用了彩色图片(即使有的书纸质版是黑白印刷的)。读者还可以方便地进行搜索、剪贴、复制和打印。

图灵社区进一步把传统出版流程与电子书出版业务紧密结合,目前已实现作译者网上交稿、编辑网上审稿、按章发布的电子出版模式。这种新的出版模式,我们称之为"敏捷出版",它可以让读者以较快的速度了解到国外最新技术图书的内容,弥补以往翻译版技术书"出版即过时"的缺憾。同时,敏捷出版使得作、译、编、读的交流更为方便,可以提前消灭书稿中的错误,最大程度地保证图书出版的质量。

优惠提示:现在购买电子书,读者将获赠书款20%的社区银子,可用于兑换纸质样书。

——最方便的开放出版平台

图灵社区向读者开放在线写作功能,协助你实现自出版和开源出版的梦想。利用"合集"功能,你就能联合二三好友共同创作一部技术参考书,以免费或收费的形式提供给读者。(收费形式须经过图灵社区立项评审。)这极大地降低了出版的门槛。只要你有写作的意愿,图灵社区就能帮助你实现这个梦想。成熟的书稿,有机会入选出版计划,同时出版纸质书。

图灵社区引进出版的外文图书,都将在立项后马上在社区公布。如果你有意翻译哪本图书,欢迎你来社区申请。只要你通过试译的考验,即可签约成为图灵的译者。当然,要想成功地完成一本书的翻译工作,是需要有坚强的毅力的。

——最直接的读者交流平台

在图灵社区,你可以十分方便地写作文章、提交勘误、发表评论,以各种方式与作译者、编辑人员和其他读者进行交流互动。提交勘误还能够获赠社区银子。

你可以积极参与社区经常开展的访谈、乐译、评选等多种活动,赢取积分和银子,积累个人声望。

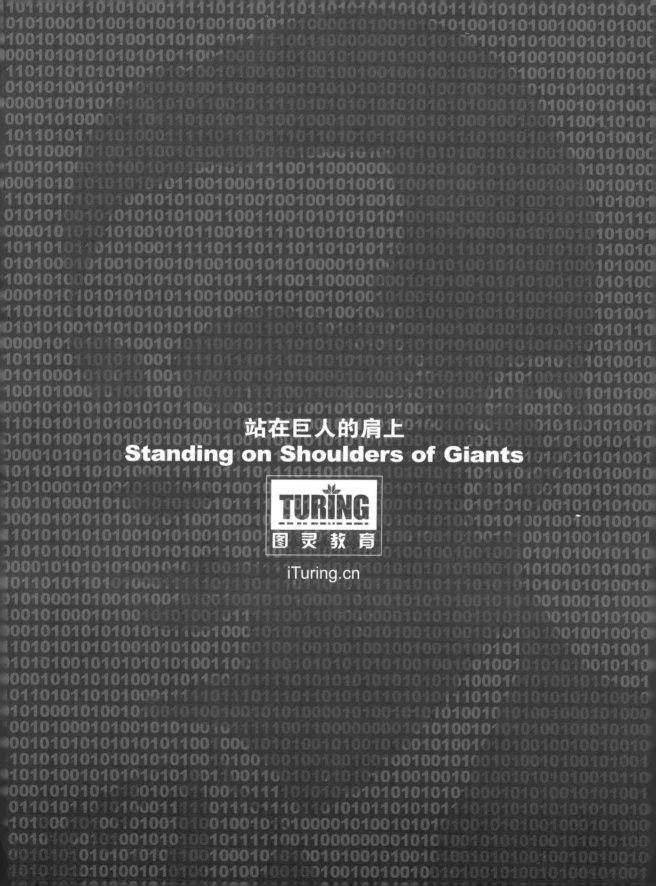